Water Basics for Decision Makers

Local Officials' Guide to Water and Wastewater Systems

By Frederick Bloetscher, PhD, PE

Published in conjunction with the Florida Section
of the American Water Works Association

The Authoritative Resource on Safe Water®

Advocacy
Communications
Conferences
Education and Training
Science and Technology
Sections

Water Basics for Decision Makers

Disclaimer
This book is provided for informational purposes only, with the understanding that the publisher, editors, and authors are not thereby engaged in rendering engineering or other professional services. The authors, editors, and publisher make no claim as to the accuracy of the book's contents, or their applicability to any particular circumstance. The editors, authors, and publisher accept no liability to any person for the information or advice provided in this book or for loss or damages incurred by any person as a result of reliance on its contents. The reader is urged to consult with an appropriate licensed professional before taking any action or making any interpretation that is within the realm of a licensed professional practice.

AWWA Publications Manager: Gay Porter De Nileon
Technical Editor/Project Manager: Martha Ripley Gray
Copy Editors: Jan Carroll, Linda Bevard
Cover Art/Production Editors: Cheryl Armstrong/Melanie Schiff

Library of Congress Cataloging-in-Publication Data

Bloetscher, Frederick.
Water basics for decision makers : local officials' guide to water and wastewater systems / By Frederick Bloetscher.
p. cm.
Includes bibliographical references and index.
ISBN 978-1-58321-585-2
1. Water--Distribution--Management. 2. Water-supply--Management. 3. Sewage disposal--Management. 4. Infrastructure (Economics)--Management. I. Title.

TD481.B58 2008
628.1--dc22

2008021703

For my parents, Frederick and Virginia Bloetscher,
in Cuyahoga Falls, Ohio, the initial (and continuing) proofreaders
of all of the documents, and Sir Thomas Orion.

This page intentionally blank.

TABLE OF CONTENTS

LIST OF TABLES AND FIGURES

TABLES

FIGURES

PREFACE

This handbook is a completely revised and rewritten version of the *Drinking Water Handbook for Public Officials* (EPA-810-B-92-016), originally prepared and published in limited quantities by the US Environmental Protection Agency in December 1992, and its successor, published in 1993 by the American Water Works Association (AWWA), which was rewritten as a user-friendly and less technical resource for local elected officials. This new edition was produced for AWWA by the Florida Section. Not only has the information from the first two versions been updated, but the data have been expanded to be of interest to all local officials, elected and appointed, including city management, finance staff, public works personnel, engineers, and new utility managers.

Prior versions of this handbook focused on water systems, but this version expands that focus to include wastewater systems. The handbook leads the reader through the process from finding the water supply, to ultimate use by the customer, and to disposal of the treated wastewater. Photos and other graphics provide the reader, who may have limited experience with water utility systems, with a clearer view of water system facilities. In addition, this version adds a section on water and sewer utilities as enterprise funds and discusses rate concepts, rate schedule options, impact fees, and financial planning. Unfortunately, in many cases where utilities are part of local governments, water and sewer funds are transferred to balance the general fund without appropriate increases in water sewer rates, which restricts reinvestment in the utility system. In the long term, this will have negative consequences for the utility. Finally, this version discusses risk, a significant and ongoing concern whether that risk is system failure, public health impacts, or breached security.

A major shift in the industry in the past 20 years has been the increase in level of technology employed and in the expertise required to operate systems efficiently. For this reason, many water and sewer system issues require the expertise of professionals and consultants. There are two results—safer water for the customer and higher costs. However, in most communities, water, necessary to our survival, is still less expensive than cable television. Water is a bargain in the United States and Canada, and its low cost makes it available to everyone.

This page intentionally blank.

ACKNOWLEDGMENTS

The author wishes to thank Gay Porter De Nileon, Melanie Fahrenbruch, Cheryl Armstrong, and Martha Ripley Gray at AWWA; Jan Carroll and Linda Bevard, copy editors; Mindy Burke, formerly an editor at AWWA, who suggested I rewrite this handbook, for her thoughts, comments, and support; Michele Miller at FSAWWA who was interested in publishing this handbook; Albert Muniz, PE, Hazen and Sawyer, PC, Boca Raton, Fla.; Mark G. Lawson, Esq., Bryant, Miller & Olive in Tallahassee, Fla.; Michael F. Nuechterlein, Esq. and Carlton Fields, Tampa, Fla.; A. John Vogt, PhD, CPA, Richard Whisnant, Esq., and David Lawrence, Esq., with the School of Government at the University of North Carolina at Chapel Hill; Daniel E. Meeroff, PhD, assistant professor, Florida Atlantic University; David A. Chin, PhD, PE, professor, University of Miami; Patrick A. Davis, PE, Hazen and Sawyer, PC, Hollywood, Fla.; David Edson, PE, Prism Environmental, Westborough, Mass.; and Gerhardt M. Witt, PG, West Palm Beach, Fla.; for their thoughts, contributions, and support; and Whitfield R. Van Cott, PE, Dan Colabella, Patricia A. Varney, Ivan Pato, and Michael J. Sheridan, PE, for whom the author has worked and at whose facilities photographs were taken for this handbook. He also thanks his fiancee Cheryl Fox for her love and support.

This page intentionally blank.

Chapter 1

INTRODUCTION

This chapter introduces the reader to the basic goals of operating a water and/or wastewater utility. Water supply and delivery, and sewer collection and treatment, have long been provided by local governments in the United States. Several factors are at the root of this delivery format—the significant investments in infrastructure that must be made, the perception that water is a basic human service that should be available to everyone regardless of income or social status, and the use of water supplies and wastewater services as a means to attract, retain, or control growth in a given area. All three reasons provide local officials with control of their future, and their constituents with the ability to hold accountable those delivering the service.

Given public control of water and sewer systems, an informed electorate has begun to become interested in water and sewer utility system issues. Customers often want to know if there are concerns that might threaten their water system and what system managers are doing to ensure that their drinking water is safe. Beyond the traditional quality concerns, the industry is now seeing issues raised about safety, security in light of the September 11, 2001, incidents, fiscal control, and fairness among consumers.

Likewise, wastewater customers want to make sure that the wastewater service is available when they need it and that the disposal of wastes is done in a manner that protects both the public health and the ecosystem. The latter includes not only the natural system but also the ability of humans to use the ecosystem for fishing, swimming, boating, and other recreational activities. The potential for public health concerns, as evidenced in the Cuyahoga River in Cleveland, Ohio, in the 1960s, is not acceptable to today's electorate.

THE MISSION OF THE UTILITY

The primary purpose of a public water and sewer system as defined by law is to provide adequate quantities of reliable service to its customers. How that is accomplished has both regulatory and policy implications. Beyond that, many utilities use mission statements to outline their goals to the public. The following are actual examples of mission statements provided by four utility systems. Each could be interpreted to say something slightly different about the outlook and priorities of the utility management staff and its governing board:

> [Our] water and sewer utility system is created to develop safe, reliable, and financially self-sufficient potable water and sanitary wastewater systems that will meet the water and sewage needs of the residents of the service area.
>
> [This] utility will provide high quality water to our customers in amounts that meet their need and protect their health at a fair price.
>
> [The department's] goal is to provide a high level of services to our residents/customers at the lowest possible cost.
>
> [This] authority will provide our customers with high-quality water and wastewater services through responsible, sustainable, and creative stewardship of the resources and assets we manage.

In each case, there is a statement about meeting the public health goals with a water supply that is safe and reliable. Providing this service confirms the health, safety, and welfare obligation of the owners, managers, and operators of the system to meet purposes to the best of their ability.

The first mission statement intends to provide service without subsidy from other sources; that is, the customers of the system will be responsible for the costs to provide the service. This mission statement indicates that the utility may be very much involved with public opinion over quality concerns and has a user base that may be willing to pay for higher-quality water than the minimum required by the regulations.

The second statement notes fairness in price. While *fairness* is not defined, the intention could be that the utility means to be competitive with its neighbors or that fairness will be exhibited among all users. The next mission statement includes "lowest possible cost." The indication is that the primary focus of policy makers is on finances, a common thread with elected officials, which may appear to contradict public health goals.

The final mission statement clearly looks at the long-term operation of the utility, using the phrase "responsible, sustainable, and creative stewardship of the resources and assets," which indicates that the utility needs to be mindful of limited water sources and the potential for contamination problems. Cost of service does not appear to be a factor except in context of "responsible use of resources." However, it is clear that protection of water resource is a major focus, which is in keeping with the intent of the Safe Drinking Water Act (SDWA; discussed in chapter 2).

Public perception of these mission statements may create unintended expectations by the public. In each mission statement, it is obvious that some interaction with customers is ongoing. All four agencies that generated these mission statements are public entities with governing bodies that respond to an electorate. Each agency is attempting to demonstrate how it intends to meet the needs of its customers. Customers dissatisfied with the service of a water system operated by governments may exert pressure on elected officials to make changes in the operation of the utility. When the problem is being caused by incompetence or lack of leadership from the public officials responsible for the water system, the public can act to replace them at the ballot box. Public agencies also will readily replace those in leadership positions to address real or perceived problems.

In contrast to the public-sector models, customers dissatisfied with the service provided by a private or investor-owned system usually cannot obtain improved service by appealing to the company either as individuals or as a group unless the situation is serious and stockholder attention can be raised. In serious cases, a regulatory agency, such as a public service commission, may become involved, but beyond serious health and safety issues, the bottom line for private or investor-owned utilities is profit. Customer satisfaction issues can affect rate requests before a public utility commission if not corrected. Because local official control over operations of private or investor-owned utilities is limited, local governments may acquire private systems where problems persist.

HOW SERIOUS IS THE RESPONSIBILITY?

Water and wastewater utility systems are oriented toward protecting the public health. Even though the results of neglect or negligence occur infrequently, the results of same may sicken or kill people. As a result, the legal system takes these responsibilities seriously. A Canadian example, the recent Walkerton, Ontario, outbreak, illustrates the extent for which responsibility may be assigned in both the United States and Canada.

Walkerton, Ontario (O'Connor 2002)

The city of Walkerton, Ontario, is a community of 4,800 people in a predominantly rural area. The city relies on wells for its water supply. In the spring of 2000, nearly half the residents became ill with what was identified as *E. coli* O157:H7, and seven residents died of the bacteria. A formal inquiry into the matter was undertaken by Justice Dennis O'Connor over the ensuing two years. Ultimately, the investigation focused on a particular well that appeared to have a surficial connection that allowed *E. coli* bacteria from a nearby field, upon which manure had been spread, to contaminate the water system. Among many findings of the investigation, Justice O'Connor noted the following deficiencies in operating the system:

- Insufficient wellfield protection (manure field in the cone of influence of the well)
- Failure of the utility to address wellfield protection efforts as required by regulatory agencies
- Poor operations by the utility staff (the water was insufficiently disinfected)
- Misrepresentation of water quality by operators, including concealing fecal coliforms in the water

- Failure of utility commissioners to respond to 1998 provincial report of deficiencies
- Inadequate funding appropriated by city council
- Funding for routine lab services and enforcement cut by province's legislature, despite knowledge of improper utility and private lab practices and warnings of health consequences
- Inadequate inspection of treatment facility by province's Office of Environment

Much can be learned from this example, and most of the points covered in this document address those issues: the need for regulatory response (wellhead protection and water quality monitoring), the importance of staff oversight and planning and of fiscal responsibility, and the emergence of new issues (a new strain of *E. coli*). Addressing the points covered in this book should allow the utility to minimize negative public perceptions.

The Walkerton example points out that the responsibility for protecting the public health, safety, and welfare extends beyond the operator and utility director, whose priorities are compliance with regulations and ensuring an uninterrupted flow of water to customers. The example also shows that because most of the decision-making responsibility for the financial obligations and the overall management lies with the utility's elected and appointed officials, they also have a responsibility to the public to uphold.

In some areas of Australia, newspapers evaluate local official performance based on

- Increases in asset value
- Increases in fiduciary position
- Decrease or minimization of system failures or breakdowns
- No catastrophic incidents

To meet these criteria, monies must be spent on asset management to optimize repair and replacement of infrastructure, maintenance should be preventive in nature instead of reactive, and new or replacement facilities should be funded as a part of the regular budget process. This means a campaigner who promises to freeze rates would probably not be viewed as being responsible (and deserving of reelection) in Australia. Most utility systems are underfunded, and most have deferred maintenance needs that are unmet. Most have limited understanding of their asset values, life, and condition (asset management). And most struggle with maintenance, as maintenance is usually the first item cut in the budget because it is very difficult to quantify. Deferring replacement of infrastructure has perils as well, as it increases the likelihood of failure of a component. Once monies are allocated for these items, there is a responsibility to ensure that managers and operators implement these programs and projects. Each of these issues will be discussed further in the later chapters of this handbook.

Chapter 2

REGULATORY FRAMEWORK

Local elected and management officials should realize that water and wastewater utility systems are highly regulated by federal, state/provincial, and local agencies. Local public officials should be cognizant of this fact when making funding and staffing decisions, as there are two laws that define the responsibilities of a utility system and its operators, both of which are public health policy driven. The first is a regulation of the water quality delivered to the customer's tap. The other is intended to improve quality in surface waters that might later be used for drinking water supplies. In both cases, the health, safety, and welfare of the public are to be protected. These are termed *police powers* for governmental entities.

The intent of water quality regulations is to assure the public that the water they receive in their homes for drinking purposes is safe to drink. Ancient civilizations realized that certain waters were healthful while others were not, the latter being water that contained some unseen contaminant that caused death or disease. With the realization that numerous epidemics, typically typhoid and cholera, were caused by waterborne diseases, people came to realize that the quality of drinking water could not be discerned simply by looking at it, tasting it, or smelling it. However, historical records indicate that the regulations or standards for water systems were generally not applied until late in the nineteenth century. From that point until the 1940s, the basic philosophy was that water should be free of any organisms that cause disease, high quantities of minerals, and such other substances that could produce adverse physiological impacts. Since then, more stringent quality criteria have been required to protect the public health.

Discharge into surface waters is one of the oldest methods of disposing of waste, because surface waters remove the waste from the point of generation. Downstream, reduction of the waste occurs due to dilution and natural degradation processes. Given sufficient treatment prior to discharge, these mutual processes work to reduce the waste to relatively minimal levels. Failure to treat adequately overloads the natural attenuation ability of the water body, resulting in noticeable pollution. In areas that are well developed, this natural attenuation does not have enough time to occur, so treatment of the wastes is needed.

For surface discharges, the physical discharge point is usually a pipe on the side of the stream (for small systems) or a pipe into the bottom of the water body with one or more holes (or diffusers) for larger systems. The most important potential environmental effect of wastewater effluent discharge into a surface water body is a decrease in dissolved oxygen levels in the surface water downstream of the discharge due to consumption of oxygen by the oxidation of organic materials in the wastewater. (The process is defined by the Streeter-Phelps equation, which mathematically models the change in dissolved oxygen with time in the presence of organic compounds and bacteria.) For this reason, secondary treatment, or 85 percent removal of oxygen-demanding substances, is generally required prior to discharge of wastewater effluent to surface water. Other potential environmental impacts of surface discharge arise from the precipitation of metals and other heavy compounds on the bottom of the receiving water body, downstream of discharge. Heavy metal accumulation among benthic populations and the possible recycling of these metals through the food chain are subjects of current research. All of these problems contribute to degraded raw water quality for downstream water plants, thereby complicating water treatment and the protection of public health.

In the United States, the two landmark pieces of legislation are the Clean Water Act (CWA) and the Safe Drinking Water Act (SDWA). The preamble for the Clean Water Act states that "the objective of this act is to restore and maintain the chemical, physical, and biological integrity of the Nation's waters." The intent portion of the 1996 amendments to the Safe Drinking Water Act states:

> The Safe Drinking Water Act Amendments of 1996 (PL 104-182) establish a new charter for the nation's public water systems, states, and the Environmental Protection Agency in protecting the safety of drinking water.

It should be noted that in both cases, the focus is oriented toward public health, not fiscal measures. Canada and Mexico also have regulations on water quality.

The responsibilities for compliance with these regulations lie with water system owners, public officials, managers, and operators. Enforcement action for failure to meet regulations is usually directed against the responsible officials of a water system, water district, municipality, or company. These responsible officials must also answer to their constituents for these failures, especially if the cause is neglect, incompetence, or misconduct in operations. The most common legal liability results from failure to comply with specific regulations. Such violations are typically periodic violations of federal, state/provincial, and local water quality standards or reporting regulations for plant operations. There are external issues with water supplies that may affect water quality and cause failures in meeting regulatory requirements. The key to limiting this type of liability is training of operators, the provision of appropriate treatment and monitoring technology, reinvestment in infrastructure on a regular basis, and redundancy in treatment processes.

Direct, personal legal liability can result from negligence and misconduct in the management and/or operation of utility systems if the public health is impacted or regulations are violated. *Negligence* is defined as the failure to exercise due care in the performance of the work, or something that an ordinarily prudent person would foresee as a risk of harm to others if not corrected. Negligence is differentiated from *incompetence*, which is the lack of ability or qualification to perform, and *misconduct*, which is a willful act that carries with it the potential for criminal penalties. Managers, officials, and employees of a water system may be exposed to a civil suit for damages if the negligent operation of the system results in injury or property damage or misconduct is noted. Although municipalities were once considered immune from suits (under the guise of sovereign immunity), the use of insurance by local governments has permitted the courts to be more sympathetic to claims where the cause can be shown to be the result of misconduct, negligence, or failure to employ generally accepted operating practices.

In the United States, much of the enforcement of the federal laws was delegated to the states during the Reagan administration. State laws often place direct responsibility on the owners and/or operators of a public water system to ensure that all laws are adhered to and good operational practices are followed.

Canada has negotiated a set of guidelines with the provinces to ensure that drinking water meets water quality criteria outlined in the *Guidelines for Canadian Drinking Water Quality*. Each province is responsible for the formulation of guidelines within the province, but the provinces typically adopt the Canadian guidelines with only minor modifications, just as the states do in the United States. Alberta and Quebec have made parts of the *Guidelines for Canadian Drinking Water Quality* into laws (which in turn makes them regulations rather than guidelines). In these provinces, it can be a criminal offense to distribute water that does not meet regulations. Most industrialized countries in the world regulate drinking water in this manner.

In Mexico, water quality and use are federal properties. Water use has been historically tied to the principle that water resources are the property of the nation and are therefore a free, constitutional right of all citizens. This is similar to some eastern states in the United States. Mexican water quality laws have been approved in the past 15 years. Laws on the management of water supplies and the use of cost-benefit analyses in the application of regulatory standards have also been approved.

REGULATORY FUNCTIONS

The major functions performed by regulatory agencies are

- Monitoring and tracking water quality data and monthly operating reports
- Sanitary surveys of water plants
- Plan review of improvements to the water system
- Technical assistance

- Labratory services
- Enforcement

Each is discussed more fully in the following paragraphs.

Monitoring and Tracking

All public water systems must monitor water quality. On-site testing of the chlorine residual in the water is performed to ensure the system retains disinfecting properties. Ongoing bacterial samples are taken for the same purpose, but they provide results 24 to 48 hours after the sample is taken. Both must be reported to the state agency with designated authority to oversee the drinking water program for the US Environmental Protection Agency (USEPA). Periodic reports of the primary and secondary contaminant analyses also must be furnished to the state agency and distributed to the customers in the form of the annual Consumer Confidence Report (CCR) in the United States or a water quality report in other jurisdictions.

The analyses for the presence of disease-causing organisms and contaminants must be performed by a certified laboratory. In some instances, this may be the water system's laboratory. A state employee may collect the samples, or the state will instruct the water system operator on how and when samples should be collected and shipped to a laboratory for analysis.

States also require periodic reports on the operation of many water systems. The system operator must provide information such as the amount of water furnished to the public, details of treatment provided, and types and quantities of chemicals added to the water. The state staff then reviews, records, and analyzes this information as one means of ensuring proper operation. Operating reports are typically submitted monthly.

Sanitary Surveys

A sanitary survey is an on-site inspection of a water system's facilities and operation. The survey is usually performed by a state employee. Survey visits range from yearly to once every several years, depending on the water source, treatment process, and resources available to the state. A sanitary survey usually involves a review of operating methods and records and a physical inspection of facilities and equipment. The sanitary survey is designed to note problems or deficiencies that could cause contamination of the water supply or interrupt continuity of service. Surveys also produce recommendations on needed programs and changes to improve water quantity, quality, and reliability. The observations and directives made during the survey are usually furnished in writing to the water system owner or operator. The expectation is that the deficiencies must be corrected within a reasonable time period or the utility will be in violation of specific regulations and subject to enforcement action.

Plan Review

The states must review all plans for construction of additions, modifications, or extensions to the water system. All plans and specifications must be prepared by a professional engineer and submitted to the regulatory agency for approval.

Such approval must be secured before construction begins. Permits are typically the vehicle to indicate approval of plans. Often included in plan reviews is an evaluation of capacity to serve the proposed extension. If capacity is not sufficient to provide the requested additional service, the permits are likely to be denied.

Technical Assistance

One of the functions of the staff of the regulatory and enforcement program is to provide system owners and operators with technical assistance. Typically, the staff members providing guidance are engineers or operators with significant experience that can come to the treatment plant to help on-site water system staff solve problems. Some professional agencies such as the American Water Works Association (AWWA) and National Rural Water Association (NRWA) have circuit riders who can provide this assistance as well.

Laboratory Services

Chemical and microbiological testing of water must be performed by a state- or USEPA-certified laboratory. Some laboratories are certified to perform a wide variety of tests while others specialize only in certain analyses. Many states/provinces have their own agency-operated laboratories that are available to perform drinking water tests, especially for small utility systems that lack resources for extensive lab work. Some agencies limit the number of samples that will be processed for a particular water system in the state laboratory, so additional required samples must be analyzed at a commercial laboratory. It is important that the quality of the commercial laboratory be determined prior to submission of samples, because incorrect or improper results may subject the water system to significant enforcement actions and adverse publicity.

Enforcement

Most violations of federal requirements are due to failure by the system operator to properly monitor water quality. Such failures subject the water system to enforcement actions. Enforcement may result in substantial civil fines against a water system owner where violations are not promptly corrected and in criminal prosecution against the individual operators. Possibly more importantly to elected officials, enforcement actions may be seen by the public as a violation of the public trust, which raises a significant public relations problem for the water system.

SAFETY RULES

In addition to regulatory requirements for water quality, monitoring, operations, and reporting, water systems must comply with safety rules. Because operations personnel come into contact with construction, roadways, machinery, chemicals, and confined spaces, operations personnel should be cognizant of the need for and proper use of hard hats, safety vests, eye protection, hearing protection, safety gloves, and other safety measures. Respiration devices are required for entry into certain confined spaces. Confined-space entry also requires specific entry training, multiple-person crews, and air-space testing.

A number of laws address safety concerns in the United States. The Occupational Safety and Health Administration (OSHA) enforces safety laws in the workplace. The Resource Conservation and Recovery Act (RCRA) regulates hazardous waste production, transportation, storage, treatment, and disposal. Many chemicals used by water treatment plants must be stored, handled, and disposed of in accordance with RCRA requirements. Hazardous material safety data sheets (MSDS) and regular training of staff in the proper use, cleanup, and emergency-leak response to chemicals is also required. Chlorine is the most dangerous chemical most water systems use. Gaseous forms may require hazard assessments. Annual training is required. Fortunately, chlorine safety training is often offered through chlorine dealers, so it is easily acquired. Safety issues should be the responsibility of a designated employee within larger utility systems.

Other North American countries also have health and safety rules that must be followed in the workplace. On-site personnel should be trained and rules enforced by supervisors in all utility organizations.

DEALING WITH REGULATORY AGENCIES

Having a series of rules and regulations to follow, regardless of how voluminous they may be, is helpful, but does not address the day-to-day issues of operating a utility system. Water system operations are variable by their very nature. Breakdowns, leaks, system damage, or failures of facilities can rarely be predicted, and each can affect the ability of the water system to meet regulatory requirements. As a result, understanding the regulations is not the only need; one must understand the regulatory perspective as well.

Regulatory agencies serve many distinct groups, as shown in Figure 2-1. Water utilities are but one of many groups that include environmental interest groups, agricultural interests, elected officials, lobbyists, and others. Each

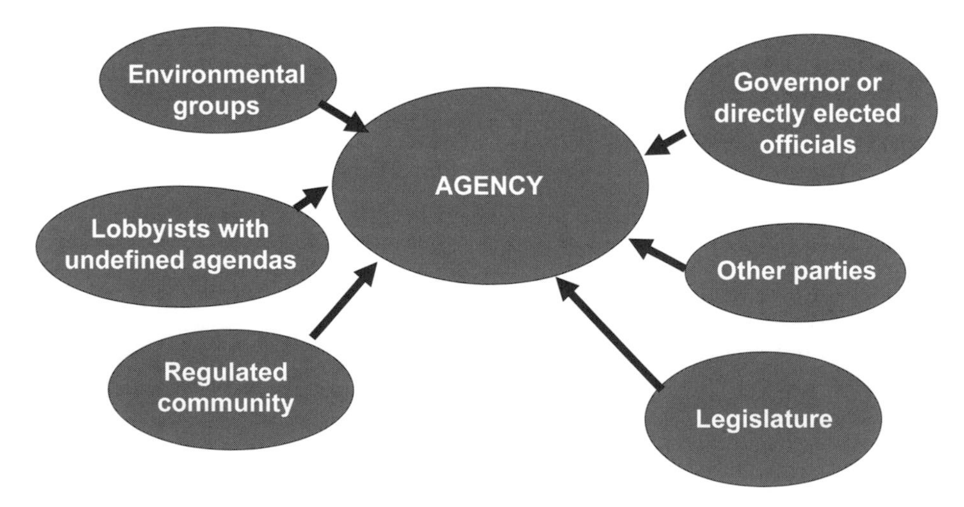

Figure 2-1 Groups providing input to regulatory agencies

group tends to have different objectives that may at times be mutually exclusive or incompatible. As a result, agency personnel need to consider all perspectives.

For utilities to work through the regulatory maze, several steps must be taken. First, the water system needs to cultivate relationships at the decision-making level. Such a relationship needs to be supportive of the agency—combatants rarely get what they want quickly. Also the relationship needs to extend before and beyond times when utility requests are considered; this is a real relationship to foster support and trust, not a relationship just to obtain a permit when desired.

Next, the water system, as a part of its ongoing efforts, needs to determine what the goals of the agency are in permitting matters, so the water system can orient its efforts to meet these goals. The goals of the agency should be demonstrated in any submittal or request by the utility. Some goals and policies may be detrimental to the utility. Therefore, the time to deal with policy issues is during the rulemaking process, not after.

This page intentionally blank.

Chapter 3

WATER AND SEWER USE AND CLASSIFICATION

Water is perpetually recycled on Earth. The recycling goes back in time far before the advent of the dinosaurs. Through what is known as the hydrologic cycle, water on Earth is continuously replenished and redistributed through precipitation, runoff, and percolation into the soil, and evaporation, condensation, and ultimately precipitation again (see Figure 3-1). The key for a water system is to determine how the hydrologic cycle provides water to the service area, in what quantities, and with what reliability. Figure 3-2 shows the variations in average rainfall across the United States.

WATER QUANTITY

The goal of a public water system is to be capable of meeting all customer quantity demands under all conditions. Therefore, the water system managers must understand the demands/usage patterns for their service area. Water use by customers generally falls into several categories.

Domestic Use

Domestic use is the consumption by residential units (private homes, condominiums, apartments, and so on). The domestic flow patterns tend to vary over the course of the day according to the time of day, the day of the week, and the time of year. On a normal day, there is higher usage in the early morning and evenings and relatively low water use through the night.

Commercial Use

Commercial use refers to the water used by stores, offices, and other businesses. A few businesses, such as laundries, restaurants, and greenhouses, use large quantities of water. Depending on the local economy and the hours businesses are open, off-hour usage may be minimal, especially late at night.

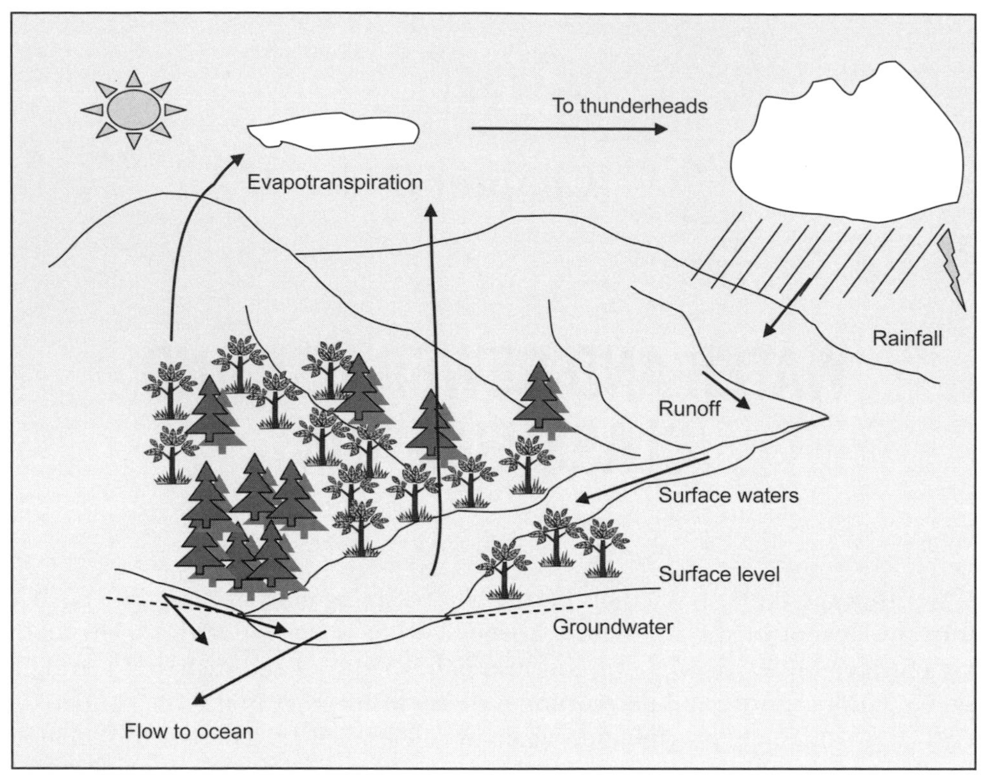

Figure 3-1 **Hydrologic cycle**

Industrial Use

Industrial use is the water used by factories and industries. A few industries use great quantities of water for cooling, cleaning, or incorporation into the product that is being manufactured, but most industries only use water for drinking, sewage disposal, and cleaning. Industries frequently have fire sprinkler systems, which require a water service adequate for drawing large quantities of water in the event of fire. Industrial use is usually predictable but varies with the number of shifts worked and days of the week the industry is in operation. In some cases, industries have wells or surface water sources for use in the plant processes and only use water from the public water system for drinking and sanitary purposes.

Urban Irrigation Use

Urban irrigation use is the water used for lawn and garden care. While billing for irrigation water is usually a part of the charge to domestic, commercial, and industrial customers, this use requires special considerations by most water systems because of the potential for huge demands in dry weather.

Irrigation demands are typically high at night. If the water system relies on nights to fill tanks, irrigation use may frustrate this effort, leaving the water system short of water in the morning. Arid and hot regions have high irrigation demands, generally in the summer, although Florida's highest demands

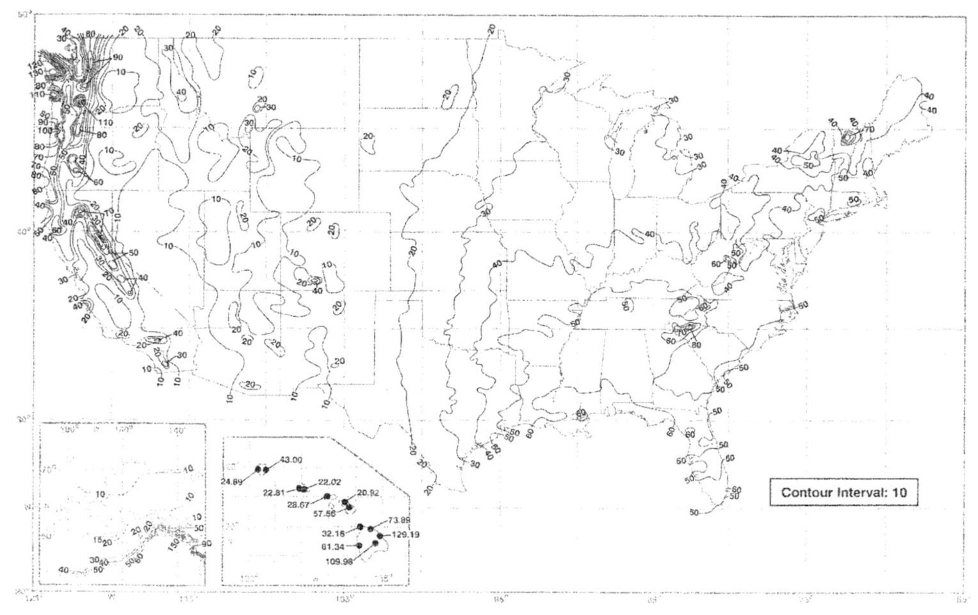

Source: NOAA, 2002

Figure 3-2 Annual precipitation across the United States

tend to be in the winter and spring due to summer rainfall patterns. Irrigation use is typically the first target for restrictions when water supplies are limited.

Agricultural Use

Agricultural use of water from a public water system is not usually practical, except where small quantities are required. In rural areas where no other water source is available, water from a public system may be used for irrigating small areas of crops and fruit trees and providing water for livestock. The same issues apply to agricultural demands as to urban irrigation demands.

Fire Use

Fire use capability is the ability to furnish adequate amounts of water for fighting fires. The amount of water required for fire protection must be in addition to water for domestic, commercial, and industrial use. Pumps are used to deliver the required demands by detecting certain minimum and average pressures throughout the system. Fire flows add demands to the system that are normally not required. The ability to furnish adequate amounts of water in the event of a fire is achieved by maintaining additional pumps and wells or reserving treated water in storage. A minimum of 20 pounds per square inch (psi) pressure is required on all drinking water pipelines.

Sewer Use

Sewage flows are generated from domestic, commercial, and industrial users. The flows are typically less than the water demands, albeit the difference may be minimal in many jurisdictions. However, most sewer systems are operated as gravity systems. As a result, if the piping is not adequately maintained, ground and surface waters will enter the pipelines, creating significant demands on the wastewater plant and possibly requiring unnecessary plant expansions. The reduction of infiltration and inflow is a major facet of operation of a sewer system and will be discussed more fully in chapter 8.

EFFECTS OF TREATMENT REQUIREMENTS ON WATER QUALITY

The quality of the raw water source is the leading factor in determining the amount of treatment to be provided and the treatment units to be constructed in a water plant. Finished-water quality must meet the regulatory requirements noted in chapter 2. The primary standards are those associated with toxins and pathogens in the water. Adverse health effects from contaminants in drinking water include many types of sickness, permanent health damage, or death.

The secondary standards regulate aesthetic qualities of drinking water, which are those parameters that may encourage consumers to use a different, and perhaps less reliable, water source. These parameters include hardness, taste, odor, color, and the tendency to discolor plumbing fixtures. Industrial and medical customers require high water quality. The water system may or may not be able to provide this quality (meaning the facility may need to treat the water further—such as is the case for a bottling company), but in either case, a reliance on a certain minimum quality is necessary. For this reason, water must be monitored and closely controlled.

SYSTEM RELIABILITY

While water quality reliance is important, water system supply reliability is absolutely essential. Losing pressure in the water distribution system for even a short time creates the possibilities of the water being contaminated by disease-causing organisms and of failure to provide appropriate fire flows. Following a loss of pressure, regulatory agencies usually require the water system to issue an order to the public to boil their water as a precaution until the distribution system can be flushed, sterilized, and tested as being safe.

The failure of treatment equipment while a water system is in operation also presents an opportunity for disease-causing organisms to contaminate the system. Adequate controls must be provided to make sure that water furnished to the public is properly treated at all times. This is why redundancy is needed for water system equipment—failure of the system is not an option.

CLASSIFICATION OF WATER SYSTEMS

Public water systems are generally classified by their source water because the source water has considerable bearing on the expected raw water quality, the required treatment, and the amount of water available for distribution to the public. Groundwater sources fall into the general categories of wells, springs, and infiltration galleries. Groundwater is not as susceptible as surface water supplies to contamination by disease-causing microorganisms, but contamination by surface activities via volatile, metallic, or synthetic chemicals is common. The most common contaminants to be removed are hardness and dissolved solids, which are typically the result of easily dissolved rock formations. Many highly productive aquifers are found in these easily dissolved and fractured formations. Both dissolved and fractured formations may have connections to surface waters or runoff, which bring contamination into the aquifer. High-quality water sources are more likely to be lower-producing aquifers, which is why most large systems depend on surface water supplies to obtain high-quality raw water supplies.

Surface water systems include water systems that draw their water from lakes and dammed reservoirs. Even where the water from a lake or river is deemed to be quite clean, they are more susceptible to contamination than groundwater systems. Surface waters have seasonal variability of sediments (suspended solids) that require substantial amounts of treatment and operational supervision. Disease-causing organisms must be removed or inactivated. Surface sources usually require a larger investment in treatment facilities and have higher operating costs as compared to groundwater sources.

Some water systems purchase their water from a water system that has already treated the water. In some instances, the purchasing system can distribute the water directly to its customers. More often, though, the purchasing utility must supply supplemental disinfection, additional storage, and repumping.

Most regulatory agencies in the United States have, in recent years, strongly recommended regionalization of water systems. There are many advantages in operating one large water system over several small ones, such as lower operating costs and improved reliability of service. The advantages are even greater if there is a high cost in obtaining the water or if the water requires expensive treatment. In some cases, regional water systems take over the responsibility for both furnishing water and operating the distribution system in all of the member communities. More discussion of this concept is included in chapter 9.

This page intentionally blank.

Chapter 4

RAW GROUNDWATER AND SURFACE WATER SUPPLIES

This chapter introduces the reader to the two major water supplies available—surface supplies and groundwater. Since the majority of utilities, especially small- and medium-size utilities, use groundwater supplies, more attention will be paid to groundwater issues. Operation of surface water systems can be complex and require more sophisticated, ongoing monitoring and protection efforts than groundwater supplies do.

SOURCES OF GROUNDWATER

The quantity and quality of groundwater depends on factors such as depth, rainfall, and geology. Throughout North America, groundwater can generally be found at some depth below the surface, from a few feet to hundreds of feet (meters). However, deeper waters tend to have poorer water quality as a result of having been in contact with the rock formation longer and dissolving of the minerals in the rock into the water. Therefore, while some deep aquifers may be prolific, the quality of water obtained from a well may not be desirable or even usable for drinking water without substantial amounts of treatment. Waters with greater than 10,000 mg/L of dissolved constituents are generally not deemed to be underground sources of drinking water (USDWs). However, it is technically possible to treat just about any water to acceptable quality, but it may not be affordable to do so. As a result, utilities generally look for least costly water sources.

In addition to quality limitations, the quantity is not always enough to meet all water supply needs. Extensive clay layers, such as exist throughout much of the southern United States, preclude significant water production beyond what is necessary for individual houses. Aquifer parameters indicate the likelihood that sufficient water can be obtained from an aquifer.

Aquifers

A portion of the water that falls on the earth as rain or snow seeps into the soil and flows downward by gravity until it contacts a layer of rock or other impenetrable material. The water then moves in a general downhill direction, taking the path of least resistance to the flow. Therefore, if underground conduits or channels are present, the water tends to flow in these pathways. The layer of soil, sand, gravel, or rock through which the water moves is called the aquifer. Water that is located near the surface, which has no rock formation above it, is called the water table aquifer (Figure 4-1). In general, the level of the water table follows the surface of the ground above, although some conditions cause exceptions. Water table aquifers are susceptible to contamination as a result of not having a rock layer above them. Aquifers located only a few feet (meters) below the surface may also be impacted by evaporation and recharge creeks, lakes, or rivers whose bottoms are deeper than the top of the water tables.

Aquifers that are located deeper and have rock formations above them are called confined aquifers. These aquifers may be under pressure. If so, they are termed artesian aquifers. Flowing artesian aquifers discharge above ground. An example is the Floridan aquifer in southeast Florida, which "flows" 30 feet (90 meters) above the ground surface. A spring forms when groundwater flows naturally from an aquifer to the land surface. Water flowing from a spring may travel hundreds of miles (kilometers) from where it seeped into the ground or could be from a source only a few yards (meters) away.

Groundwater dissolves minerals in the rock formation. The movement of water through an aquifer is generally quite slow, but eventually the large cavities can interconnect and form underground rivers or caverns that can be tapped as a public water supply source. Some aquifers are buried river valleys or the beds of an ancient lake. Water may travel 20 feet (6 meters) or more per day

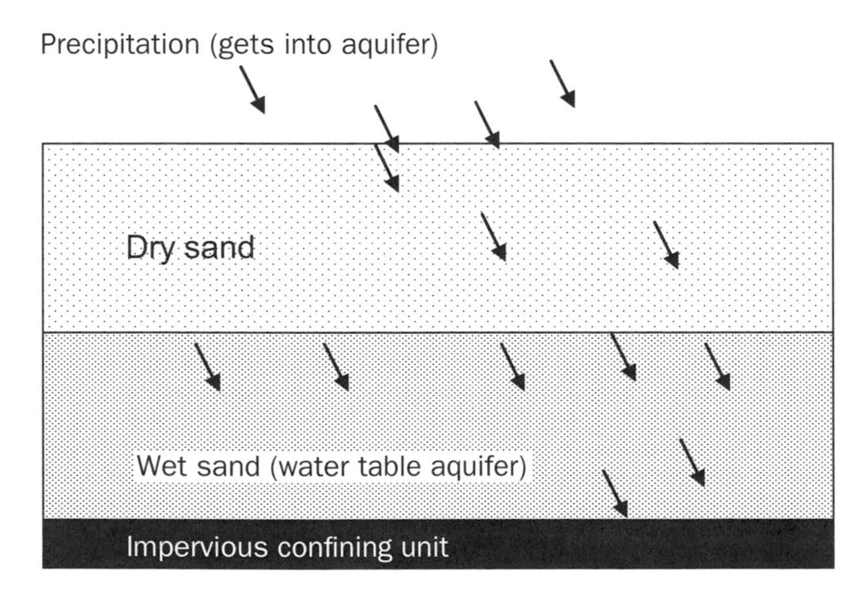

Figure 4-1 Water table aquifer

Table 4-1 Porosity of rock formations

Strata	Porosity (%)	Spec Yield (%)	Specific Retention
Soil	55	40	15
Clay	50	2	48
Sand	25	22	3
Gravel	20	19	1
Limestone	20	18	2
Sandstone	11	6	5
Granite	0.1	0.09	0.01
Basalt	11	8	3

in coarse sand whereas in fine sandstone, it may move only a few feet (1 meter or less) in one year and not at all in clay. The flow velocity and flow direction of groundwater depends on the permeability of soil and rock layers, and the relative pressure of groundwater. Sand and gravel aquifers are usually the most suitable for public water system use because their relatively high permeability provides significant productivity. Sandstone is porous and often yields water of good quality in sufficient quantity to supply public water systems. Limestone is not very porous but often has cracks and cavities that can provide sufficient quantities of water. Table 4-1 shows the porosity of rock formations.

Where there are several aquifers, the aquifer layers are separated by a layer of material such as clay (a confining unit or aquitard). A public utility must evaluate the productivity and water quality of each aquifer because the quantity and quality of water in the aquifers may vary greatly. This can be done with aquifer testing, review of existing geological reports, and other methods noted later in this chapter.

Gravity and Artesian Wells

A gravity well is a hole or shaft dug or drilled from the ground surface into a water table. Water must then be pumped from the water table to the surface for use. The operation of gravity wells tends to create a cone-shaped depression (in three dimensions) in the groundwater surrounding the well. This is called the *cone of depression,* and the distance that the water level is lowered in the well during pumping is called the *drawdown* (see Figure 4-2). The drawdown amount is important when selecting pumps, monitoring water supplies, and checking pump or well performance.

The diameter of the cone of depression and the drawdown varies with the size of the well, pumping rate, and the water flow rate through the aquifer. In porous sand and gravel, the depression may be small. In denser soils, it is very significant. Wells are normally placed far enough apart so that their zones of depression do not overlap significantly. Wells should be sized to minimize

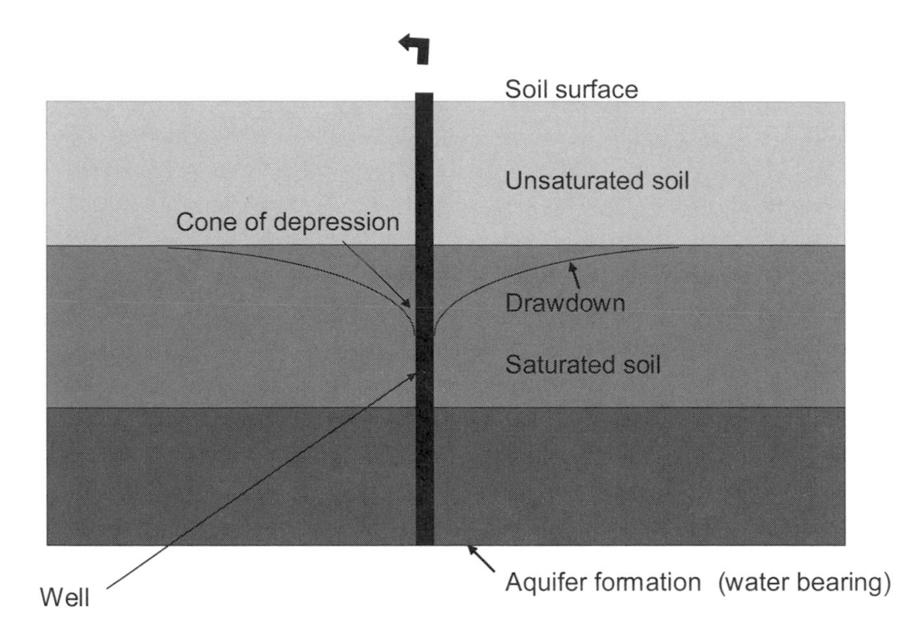

Figure 4-2 Well drawdown

drawdown so that the aquifer can rebound quickly. Permanent drops in water levels can occur when the aquifer is pumped too much. The Black Creek aquifer in eastern North Carolina is such an example. When water cannot recharge the aquifer, the water system will need to find a new source.

Ideally, wells should pump continuously without permanent drawdowns occurring. Where this is not possible, it is common practice to pump wells that have a significant drawdown for only a few hours each day, allowing an extended period for the aquifer to recover. The long-term capacity of an aquifer can be determined by tests and study of the geology of the area. Water system managers who are unsure of the amount of water available should consult state geological experts or professional hydrogeological firms.

An artesian well is constructed to tap an aquifer that is located under a confining layer where the water is under some pressure. Where the pressure is sufficient for the water to flow to the ground surface, pumping may not be needed. After an artesian well has been used for a period of time, the pressure reduces until the water no longer flows to the surface. A well is still called artesian if the water level rises above the top of the aquifer.

GROUNDWATER PROTECTION

Water table aquifers can be contaminated easily by spills from surface activities. A common problem has been leakage from underground gas storage tanks, which has contaminated hundreds of sites across the United States. Under the Superfund Amendments and Reauthorization Act (SARA) program, the US Environmental Protection Agency (USEPA) prioritized cleanup sites. Because of their small volume, gas stations have not always been high on the priority list for USEPA. Some states have made them a higher priority, but the number of sites and limitations in funding hurt cleanup efforts. Once cleanup is needed, it is a long process. A better practice is to locate wellfields where contamination is unlikely, then protect the wellfield from surficial impacts. This is the focus of wellhead protection programs that are a part of the Safe Drinking Water Act (SDWA) amendments. As noted in chapter 2, in the 1960s, USEPA found that nearly a quarter of wellfields were impacted by surface contaminants. Improved chemical analysis and more complete sampling of well water have revealed a large number of public water supply wells contaminated by careless use and disposal of synthetic chemicals.

States and provinces are now implementing regulations to protect underground sources of water. Some actions being taken include

- New requirements for installation and testing of underground storage tanks
- Increased regulation for handling, using, and transporting toxic chemicals to reduce the possibility of spills
- Greatly increased regulation of landfills and other waste disposal sites
- Closer control of the use of pesticides and agricultural chemicals, sampling, and monitoring of identified groundwater contamination locations
- Action to remove the contamination

All of these programs will affect land use, possibly affecting local constituents, especially where private property rights laws are an issue. However, modeling the wellfields with sophisticated computer models will allow the water system to develop areas where certain activities should be limited. Figure 4-3 is an example of the wellfield protection areas from Broward County, Fla. The county has four tiers or zones for regulation based on travel times from the surface to existing wells. Activities are limited based on these tiers/zones (Tier/Zone 4 has no requirements).

In conjunction with wellhead protection efforts, water systems should identify any groundwater sources they are using that may be directly affected by surface water. The concern is that if there is minimal filtration that occurs between the surface and the water withdrawn for wells, contaminants, especially microbiological constituents, may contaminate the water source. Water sources that meet these criteria are considered groundwater under the direct influence of surface water (GWUDI). If an aquifer is determined to be groundwater under the direct influence of surface water, and therefore vulnerable to contamination by disease-causing organisms found in surface water, the well water must be treated under the same requirements as for a surface water system, meaning mandatory disinfection and filtration.

WELL CONSTRUCTION AND OPERATION

Siting Considerations

Well siting considerations include four issues: site availability, water supply, water quality, and wellhead protection limitations. Many small water systems have

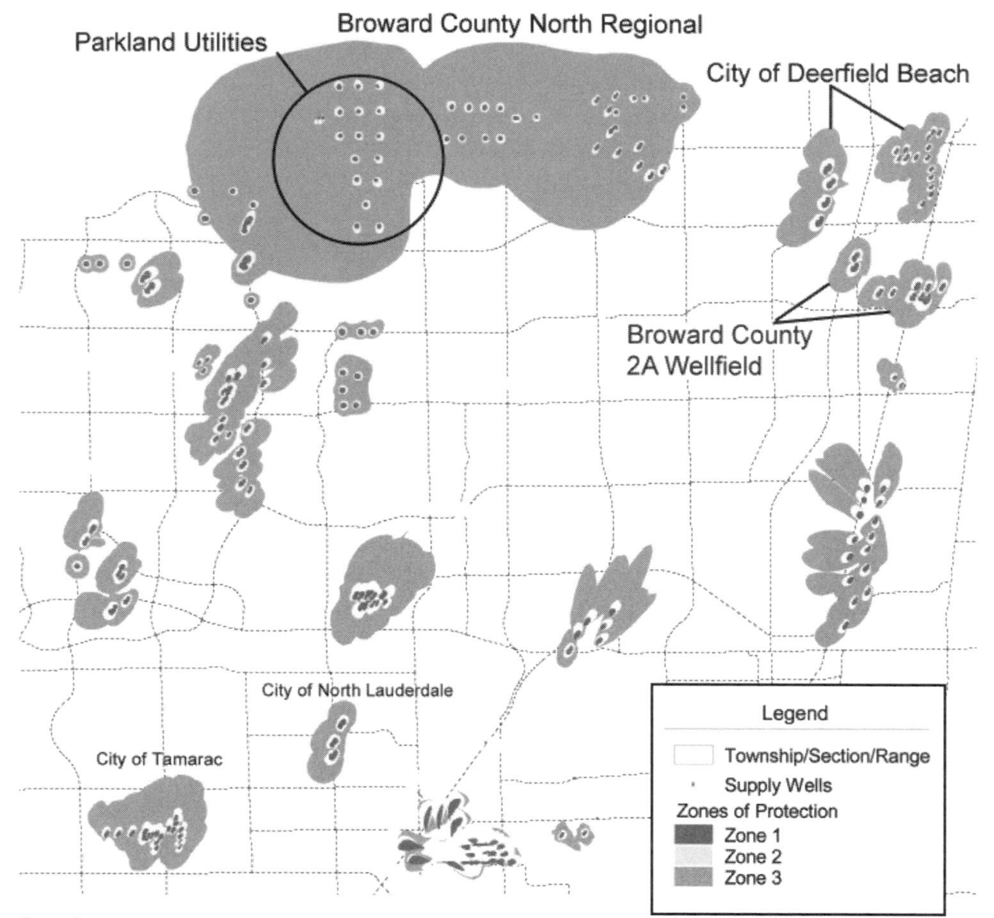

Zone 1:

Provides up to a 10-day buffer around the wellfield. No hazardous chemicals (regulated substances) are permitted within Zone 1.

Zone 2:

Provides up to a 30-day buffer. Businesses are required to be licensed and test the groundwater at their facility for regulated substances they store or use on site.

Zone 3:

Provides up to a 210-day buffer. Businesses are required to be licensed and secondary containment is mandated for their stored regulated substances.

Source: Broward County 2001.

Figure 4-3 Broward County wellhead protection zones

made the prime considerations in selecting sites for public water supply wells the cost of the land. As a result, many old water system wells are found next to city halls, police stations, or other government buildings. This generally violates wellhead protection efforts. While land may be cheap, using only land cost as the criteria may complicate water quality and protection efforts.

Water quantity and quality are intertwined. The water system must balance factors including well depth, geology of the area, characteristics of the rock formations, and dissolved minerals in the aquifer. In most places, locating productive wells may be accomplished using local knowledge of the geology of the region and the experience with existing wells. Most states have an agency that keeps geological records and can furnish information on likely aquifers. Good information on the experience with wells in an area is usually available from local well-drilling firms and hydrogeologists.

Local knowledge and a hydrogeologist can determine whether productive layers exist. However, there are no guarantees. To optimize water production, the water system will need to employ geophysical methods along with test wells to specifically locate productive aquifers. Geophysics is especially important when considering deep, high-capacity wells. Firms specializing in geophysics use scientific instruments to determine potential water-bearing strata from a test well. Figure 4-4 is an example of some geophysical logging results. Multiple log types are required to render a conclusion about the aquifer. The significance of the logs should be interpreted by a professional hydrogeologist.

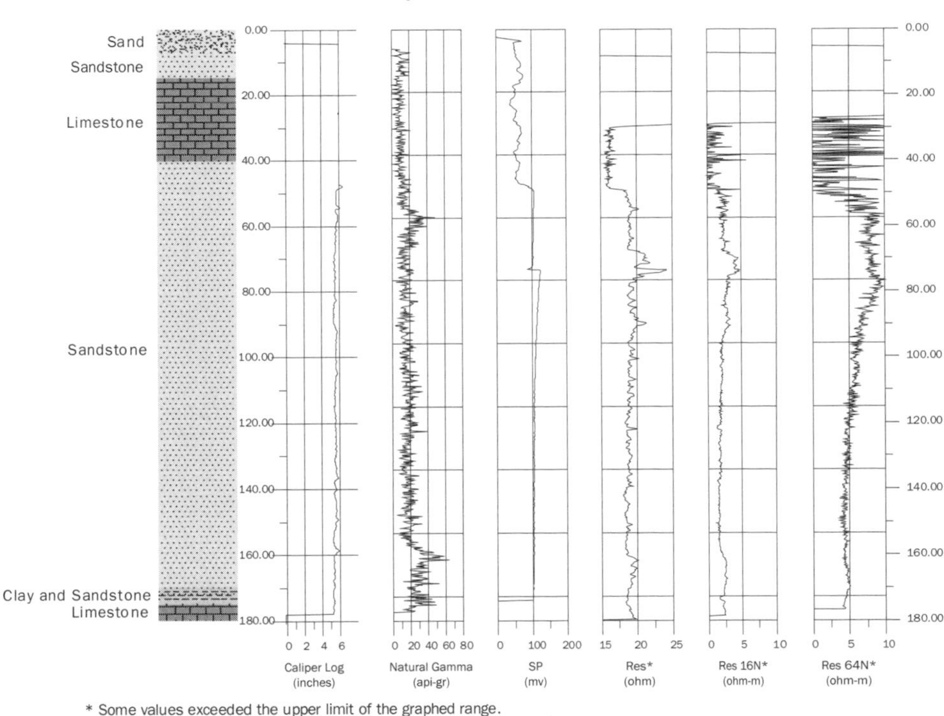

Source: Bloetscher et al. 2005.

Figure 4-4 Geophysical logging example

Wells should be modeled by professionals to determine appropriate spacing to limit drawdown of the aquifer. When an aquifer is very productive and wells have small zones of influence, wells may be located close together in a wellfield. Figure 4-5 shows the drawdown for the Fort Lauderdale Peele Dixie wellfield. Having the wells in a group has the advantage of simplified monitoring and maintenance and allows water from all of them to be treated at a single plant before it is pumped to the distribution system. Well construction should be standardized to simplify maintenance efforts.

Two other factors that impact well siting are flooding and water rights. Flood protection is required to prevent possible contamination by surface water during a flood and to guard against water damage to the mechanical and electrical equipment. State laws generally prohibit construction of wells in a flood zone. Water rights laws are a major issue in the western 18 states but vary in the eastern states. These laws may restrict the amount of water allowed to be withdrawn, the timing, and the locations where a public water system can withdraw water. These policies may have considerable bearing on the siting of new wells and should be carefully reviewed before contracting for new well construction.

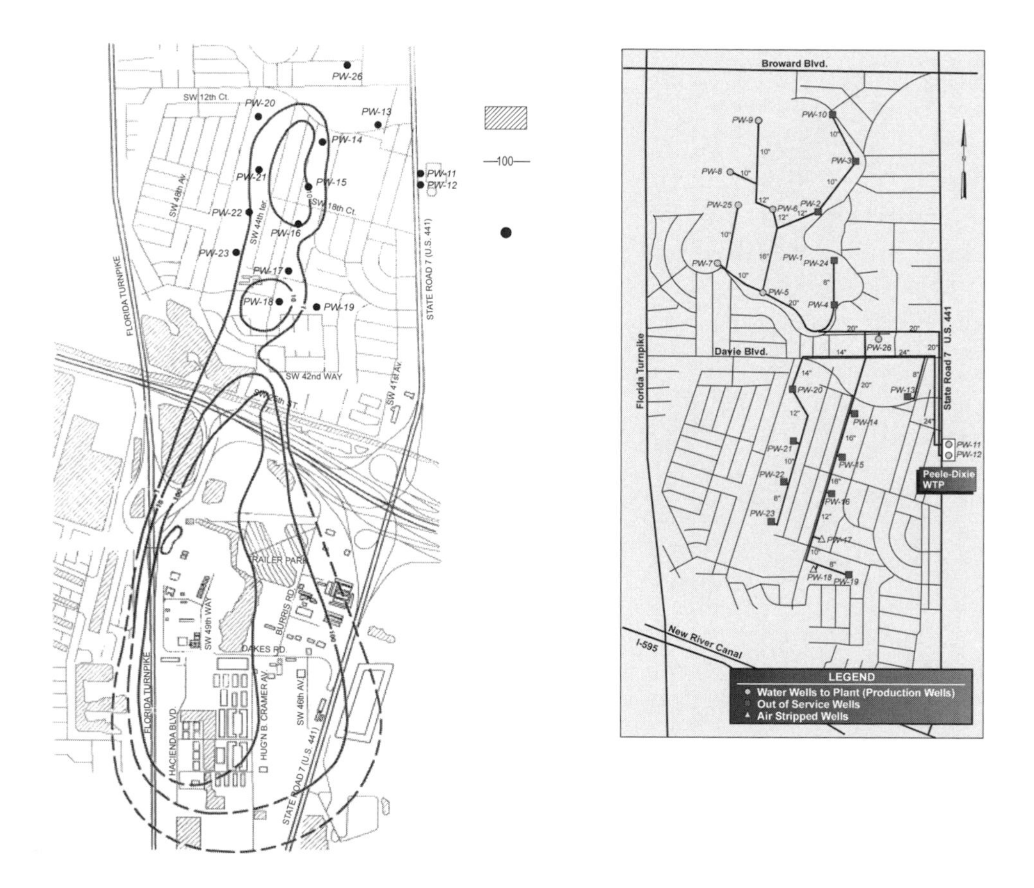

Source: Hazen and Sawyer, P. C., Boca Raton, Fla.

Figure 4-5 Example of wellfield modeling drawdown contours—Fort Lauderdale Peele Dixie wellfield

Well Construction

Principal well construction types include dug, jetted, bored, driven, and drilled wells. Most wells serving community public water supplies are drilled. Drilled wells are used where a larger-diameter pipe is needed, where hard ground or rock may be encountered, or where the well must be deep. Drilling requires the use of a derrick or crane to hold the drilling tools. The drilling bit is forced through the earth and rock by being repeatedly lifted and dropped on the end of a cable or by being rotated on the end of a shaft. Drilled wells are most commonly installed for public water systems because it is often necessary to tap deeper aquifers to obtain adequate quantity or obtain the best water quality. Drilled wells can also be constructed with a large diameter for installation of large-capacity pumps. Figure 4-6 shows an example of a well design.

Monitoring

Two issues must be monitored in a wellfield on a regular basis: water quality and well performance. Good operating practice requires that well water levels be measured periodically, both while the pump is idle (static level) and while the pump is operating (drawdown level). Changes in static and drawdown levels should be periodically reviewed for trends. Two trends may be indicated: long-term reductions in water availability in the aquifer or limitations in specific capacity of the mechanical system caused by pump wear, clogged screens or formations, or bacterial fouling. Repairs to correct mechanical problems should be scheduled before they become serious. If the water table is dropping in a well, the pump will eventually have to be lowered, replaced, or used less, or more spacing will be needed between wells.

Regulations require periodic monitoring of microbiological and chemical quality. Monitoring water quality should occur initially to establish a baseline that, once established, can permit the water system to reduce the frequency of groundwater sampling. Fortunately, groundwater quality in many locations does not change, because the movement of groundwater is very slow compared with surface water. Where contamination risks are high, sentinel monitor wells should be installed. These wells, located at various depths, will provide definition for the initial groundwater assessment. Sentinel wells also serve as an early warning system to detect changes in water quality and water elevations before they affect the water supply wells.

If laboratory analyses identify contamination in a well, additional samples must be taken to confirm the initial results. If chemical contamination is found to be present, but it does not exceed the maximum contaminant level (MCL), the state usually requires increased monitoring to determine whether the concentration is increasing. If testing confirms chemical contamination at levels that exceed the MCL, use of the well must be discontinued or treatment installed to reduce the level of contamination.

If microbiological contamination is confirmed, regulatory agency representatives advise on steps that must be taken immediately to identify the source of contamination, disinfect the well, and provide required public notification. Fecal coliforms generally indicate contamination with sewage—often from septic

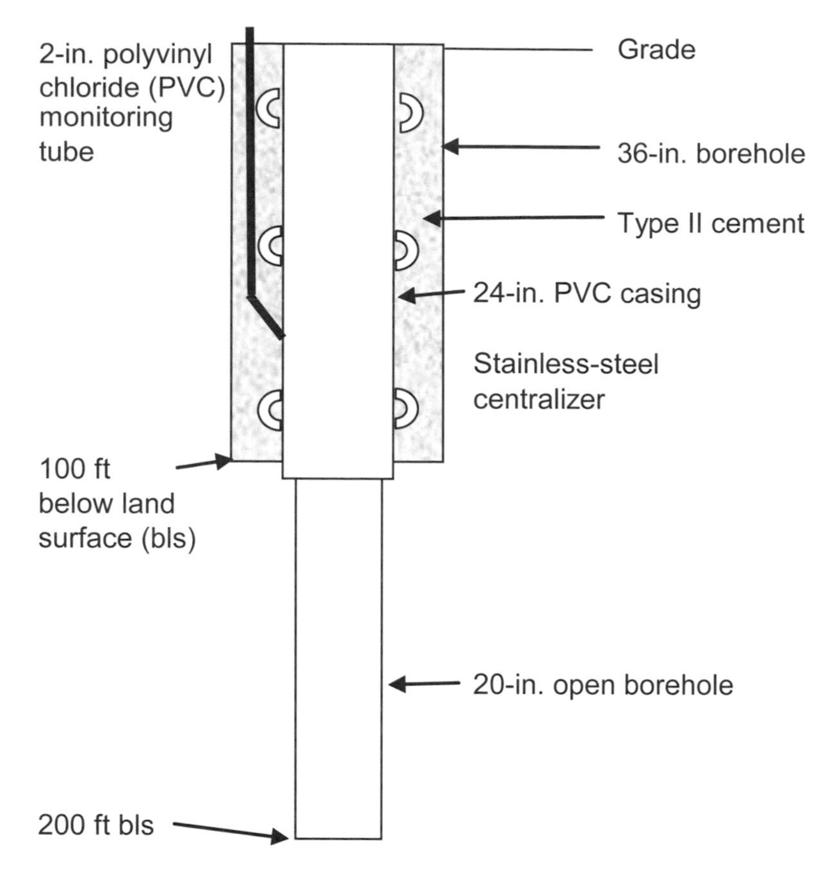

Source: Bloetscher, Muniz, and Witt 2005.

Figure 4-6 Typical well construction

tanks or farms. Fecal coliforms must be treated by disinfection. The well should be taken off-line and disinfected. The well should not be returned to service until the analyses indicate the well is free from fecal coliforms.

However, it should be noted that certain microbiological activity is normal. Aquifers are not the pristine environments the public may believe. Aquifers are ideal places for bacteria to grow because of stability in the environment and adequate supplies of organics (Bloetscher and Saltrick 1998). Bacteria in wells tend to have an impact on the life of the wells, their operation, and the integrity of the materials used to construct the wells as a result of biofouling degradation. The typical agents for microbiological fouling include iron-reducing bacteria (IRB), sulfur-reducing bacteria (SRB), and slime-forming bacteria (SFB), although many others exist. Iron bacteria, like *Gallionella*, are common in aerobic environments where iron and oxygen are present in the groundwater and where ferrous materials exist in the aquifer formation (such as steel or cast-iron wells). These bacteria attach themselves to the steel and create differentially charged points on the surface, which in turn create cathodic corrosion problems. The iron bacteria then metabolize the iron that is more soluble. Iron bacteria tend to be rust colored or produce rust-colored colonies on pipe surfaces

Figure 4-7 Bacteria on well column pipe

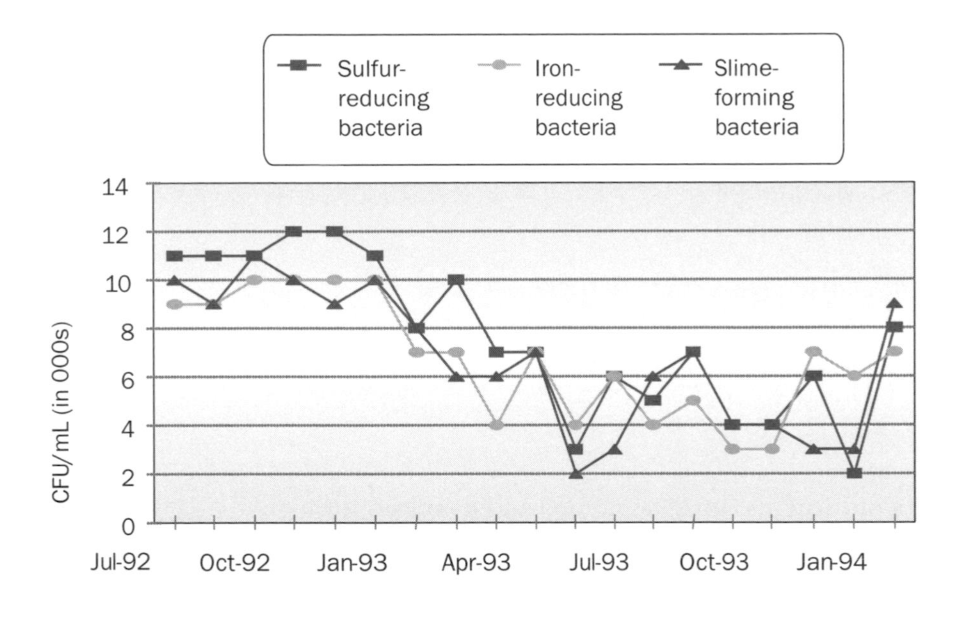

Figure 4-8 Venice bacterial samples

(see Figure 4-7). Sulfur-reducing bacteria are often responsible for the release of hydrogen sulfide when raw water is aerated. Monitoring bacteria population is important. Figure 4-8 provides monitoring data for SRB, IRB, and SFB from a well in Venice, Fla.

Owners and operators of wells that have been identified as having increasing levels of contamination should immediately begin assessing their alternatives for correcting the problem. The major choices that may be considered in attempting to locate the source of contamination are to

- See if correction or removal of the source will allow the aquifer to return to normal
- Determine whether the plume of contamination flowing toward the well can be blocked or intercepted
- Determine if it is economically feasible to treat the water to remove the contamination
- Investigate whether altering the well to draw water from a different aquifer is possible
- Investigate the feasibility of drilling a replacement well at another location where there is no contamination
- Investigate whether water from the contaminated well can be blended with water from an uncontaminated source to bring the finished-water level below the MCL
- Investigate changing to a surface water source
- Investigate purchasing water from another water system

Although groundwater quality does not vary much with time, certain quality features do gradually change. The following parameters should be analyzed in raw water on a regular basis.

- Specific ions in water
  - > Calcium (Ca^{+2})
  - > Magnesium (Mg^{+2})
  - > Sodium (Na^{+})
  - > Strontium (Sr^{+2})
  - > Barium (Ba^{+2})
  - > Fluoride (F^{-})
  - > Sulfate (SO_4^{-2})
  - > Chloride (Cl^{-})
  - > Bicarbonate (HCO_3^{-})
  - > Carbonate (CO_3^{-2})
  - > Nitrate (NO_3^{-})
  - > Potassium (K^{+})
- Constituents (total–combined and dissolved forms)
  - > Aluminum (Al^{+3}) (total and dissolved)
  - > Manganese (Mn^{+2}) (total and dissolved)
  - > Iron (Fe^{+2}) (totaled, dissolved, and ferrous)
  - > Phosphate (PO_4^{-3}) (total)
  - > Silica as silica dioxide (SiO_2) (total and dissolved)
  - > Total organic carbon (TOC)
  - > Hydrogen sulfide (H_2S)
  - > Free chlorine (Cl_2)
  - > Dissolved oxygen (O_2)
- Other water quality parameters
  - > Total dissolved solids (TDS)
  - > Carbon dioxide (CO_2)

> pH
> Temperature
> Turbidity (nephelometric method)
> Silt density index

Maintenance

Wells require periodic maintenance. One of the most common well problems is incrustation of the well screen or of the gravel pack around the screen. This may be due to release of dissolved minerals from the native water, geochemical reactions, or microbiological activity as noted in the previous section.

Calcium carbonate creates problems because it forms a scale on the screen and cements together particles of sand and gravel. Usually, a chemical process can remove calcium carbonate incrustation. In some cases, maintenance of a well must be done as often as yearly.

Sand, silt, and other particulates may clog the screen. When this happens, the capacity of the well decreases, the pumps become less efficient, and operations costs for electricity increase. Sand also increases wear on pumps and settles in large pipelines. Sand is very problematic for membrane processes. Removing sand without damaging the screen can be a delicate process. Water samples from wells developed in sand aquifers should also be periodically inspected for the presence of sand. The presence of sand in a well may be an indication of eventual collapse of the well, collapse of the formation, and in extreme cases, sinkholes. Such problems are generally repairable but require that someone with the appropriate expertise review the situation.

SURFACE WATER SOURCES

It is relatively rare to have sufficient groundwater available to serve a large community. As a result, the majority of people served from central water systems in the United States are served from surface water sources. Historically, large cities developed on the banks of large, steady lakes and rivers (such as the Mississippi River and the Great Lakes). As technology has improved, diversions of waters from long distances or across drainage basin divides have been accomplished. Southern California and Denver, Colo., are two places where interbasin transfers of large quantities of water have occurred. Rivers and lakes are the water bodies typically used for water supply because the large quantities that are available make them attractive to large water systems. Reservoirs can help retain water where variations in precipitation during the year may make them desirable.

Rivers and Lakes

Rivers. The goal of the clean water legislation is to clean up the rivers and lakes because many water systems obtain their supplies from rivers that are very polluted. Since the 1960s, significant progress has been made to clean up the rivers in the United States and Canada, but variations continue to exist, and some rivers remain quite polluted. Rivers like the Ohio and Mississippi provide water of widely varying quality and quantity. Other rivers, such as those in the Rockies, supply smaller or more seasonal amounts.

Land that has been developed may discharge treated sewage or contribute polluted runoff to adjacent or local water bodies. Stormwater runoff may include significant quantities of synthetic organics and petroleum products. Agricultural areas permit the runoff of nutrients (fertilizers), herbicides, pesticides, and animal wastes, plus hormones and antibiotics from animal feedlots. While the microorganisms in the rivers have some capacity to break down waste, many waste products remain in these polluted water bodies. These contaminants can cause undesirable tastes, odors, and color in the water as well as adverse health effects. Constant monitoring of surface water supplies for turbidity, nutrients, and bacteria is necessary to ensure water is of sufficient quality to provide to residents. Treatment plant operations must be continually adjusted to account for variations in raw water quality.

One disadvantage of surface water sources is that water quality and quantity can vary from day to day and even from season to season. Water quality in some rivers is much different in the spring as opposed to the fall because snowmelt and rains carry silt and decaying vegetation off of forest floors, fields, and urban areas. Some rivers may be in flood stage in spring and flow at a very low rate in midsummer. Treatment facilities must be capable of treating under all conditions. The result is that much more monitoring occurs with surface waters than groundwaters.

Lakes and Reservoirs. Lakes are naturally occurring bodies of water. Reservoirs are generally artificial lakes or impoundments formed by constructing a dam across a stream to collect and store water. The terms are often used interchangeably. Reservoirs may be constructed for a combination of reasons, including flood control, electric power generation, and the provision of an adequate source of water for public water supply and irrigation during periods of low stream flow such as in the Tennessee Valley Authority projects (see Figure 4-9).

The water from lakes and reservoirs is often of better quality than water from rivers because retention time allows contaminants, especially large contaminants, to settle. While solids settle, other contaminants are reduced by biological action or their concentration is reduced by dilution before the water is withdrawn for public use. However, in stagnant lakes and reservoirs, dissolved oxygen can decrease in the summer months causing plant and algae growth, which may cause tastes and odors in the water. Recreational use of lakes and reservoirs may introduce other pollutants, so care should be exercised when permitting power boating and other human activities on lakes and reservoirs so as not to reduce the water quality for drinking water purposes.

Lakes and reservoirs have one other feature that must be addressed—temperature variations in the spring and fall cause the lake strata to exchange: warm water rises, cool water descends (see Figure 4-10). When this happens, the water from a lake or reservoir may be difficult to treat due to turbidity. Arrangements to address such circumstances must be designed into surface treatment facilities.

Figure 4-9 Tennessee Valley Authority reservoir in North Carolina

Source Considerations

Source Selection. Surface water bodies are limited, so a water system typically has limited choices when selecting a source. Competing users may also challenge potable water as the best use of the water body; recreation, agriculture, and ecosystem needs may be higher priorities from an economic perspective. If a choice of sources exists, considerations have to be made between the water quantity and quality available and the cost of obtaining and treating the water from each source. In addition, considerations of potential future uses should be undertaken, since some uses may degrade water quality or reduce the quantity available. One option in such circumstances is to acquire the watershed, as Seattle and New York City have done.

Almost any water can be treated to acceptable drinking water quality with enough investment in treatment facilities and operating costs. The consideration is how reasonable the cost for treatment is. Generally, clean water has lesser treatment costs and is easier to keep clean and consistent than poor-quality source waters. It may be more cost-effective to pipe water a considerable distance from a high-quality source than to provide treatment for a poor local source.

Surface Water Contaminants. The suspended matter that causes water to appear cloudy is called turbidity. Almost all surface water has turbidity present to some degree. It is usually a combination of sand, silt, small organisms, and plant matter ranging from minute to fairly large particles. Some turbidity particles settle quickly. Others are so close to the density of water that they stay in suspension almost indefinitely. Turbidity is a problem because it provides a vehicle for organisms to use to resist disinfection. Highly turbid water cannot be disinfected adequately because the chlorine reacts with the suspended matter, not with the microorganisms. As a result, filtration of surface waters is required

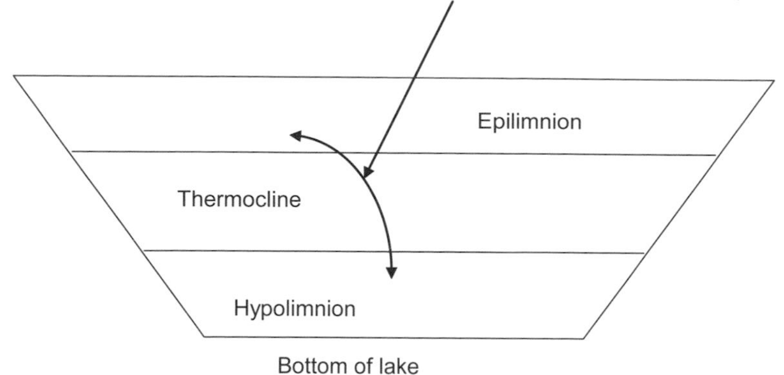

Figure 4-10 Layers of stratification in a typical lake

to remove turbidity. All surface water is therefore considered potentially contaminated with microorganisms harmful to human health.

Chemical contamination occurs from agricultural and industrial areas. Major concerns with agricultural runoff into surface waters include the presence of pesticides, herbicides, antibiotics, and endocrine disruptors. Stormwater runoff often carries large amounts of animal wastes, fertilizer, and pesticides from fields to lakes and rivers in both agricultural and developed areas. This nonpoint source of pollution often causes high levels of nutrients in the water, which can lead to increased turbidity and outbreaks of algae and nuisance plants in lakes and reservoirs. There are limited treatment options available for all of these contaminants.

Water Rights. Water rights laws in the 18 western states may restrict or limit withdrawals of water by a water system. These laws must be carefully reviewed as a first step in selecting a surface water source. Certain eastern states regulate water withdrawals in areas where water availability may be limited.

SURFACE WATER WITHDRAWALS

Surface Water Intakes

A surface water system requires an intake, just as a groundwater system requires a well. An intake structure (Figure 4-11) must be located so that it will collect the best possible water at a given time; that is, these structures usually have multiple levels from which water can be withdrawn. Intakes must be protected from damage by vandalism, ice, boats, and floods. In rare instances, or for small systems, water of acceptable quality can be drawn from the shore of a lake or river (note that the level may rise and fall with seasons and annual variations in rainfall). However, an intake located at the surface will draw water of variable temperature and quality and may become clogged by floating debris or ice in the winter months. Intakes on the bottom may pick up debris as well as fine materials. The best-quality water at any given time must be determined by testing.

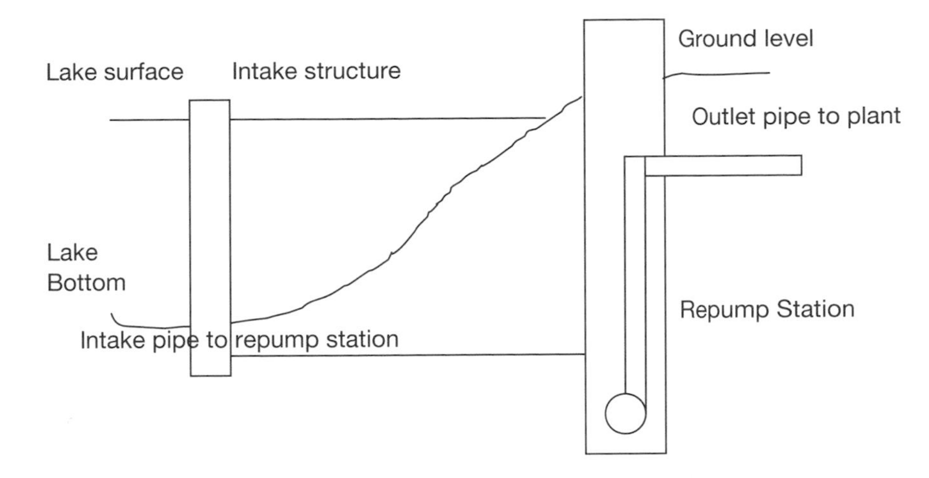

Figure 4-11 Typical surface water intake system

The pipe leading from the intake to the shore is usually trenched into the bottom of the water body to protect it, and the intake structure is raised to avoid drawing in silt from the bottom. The water drawn through an intake usually flows by gravity into an intake structure located on shore near the treatment plant. This structure helps equalize flow and provides a place for debris to settle out so that damage to pumps and treatment equipment will be minimized.

In recent years, zebra mussels and other types of freshwater mollusks have invaded the Great Lakes and other surface waters in the United States. The forecast is that they could eventually be present in most lakes and streams. These shellfish attach to underwater structures, including water intakes and pipelines. When the mollusk dies, the shell is left attached. The potential problems for water systems include blocking of the intake inlet, reduction of intake pipe flow, and tastes and odors that accompany die-off of the mollusks. The design of all new water intakes should include consideration of methods for controlling the growth of mollusks in the event that they eventually invade the water source.

This page intentionally blank.

Chapter 5

WATER TREATMENT

The intent of this chapter is to introduce the reader to the more common methods for treating water. The section is divided into groundwater and surface water treatment, as they are often treated somewhat differently. As noted in the last chapter, since the majority of utilities, especially small- and medium-size utilities, use groundwater supplies, more attention will be paid to groundwater treatment. However, surface water supplies tend to have more extensive and perhaps consistent treatment methods. Both systems tend to have filters and some form of chemical coagulant, although the function is usually different. Both have disinfection as required by the drinking water regulations.

TREATMENT FOR GROUNDWATER QUALITY ISSUES

Groundwater supplies are never *pristine* or *pure* water—there are always some constituents dissolved from the aquifer formation. Therefore, most groundwater supplies require some form of treatment. The extent of treatment is related to water quality. Aquifers that are well confined, have few dissolved minerals, have limited microbiological contamination, and generally meet the Safe Drinking Water Act (SDWA) standards will likely require only disinfection (see Figure 5-1). Where high-quality aquifers exist, treatment can be decentralized. However, if the water contains substances that may require treatment beyond disinfection or aeration, decentralized wells should not be constructed.

Most groundwater contains dissolved minerals or other natural constituents that require softening (see Figure 5-2) or ion exchange. In many formations, the dissolved minerals constitute alkalinity and/or hardness—limestone is a typical example (see the next section), but iron, sulfates, and other constituents may also exist. High hardness means medium-quality water. A significant relationship exists between mineral quality and the depth below the surface where the groundwater is withdrawn: the mineral content of groundwater generally increases with depth except along the ocean coast. In deeper formations and where seas once covered the formation, the dissolved minerals may be salts

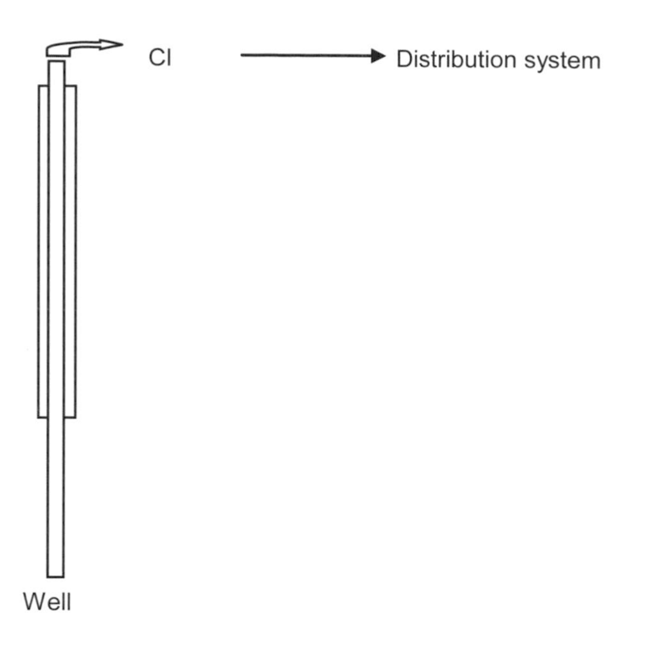

Figure 5-1 High-quality groundwater treatment regime
Raw water has low hardness, no metals, and generally meets SDWA standards.

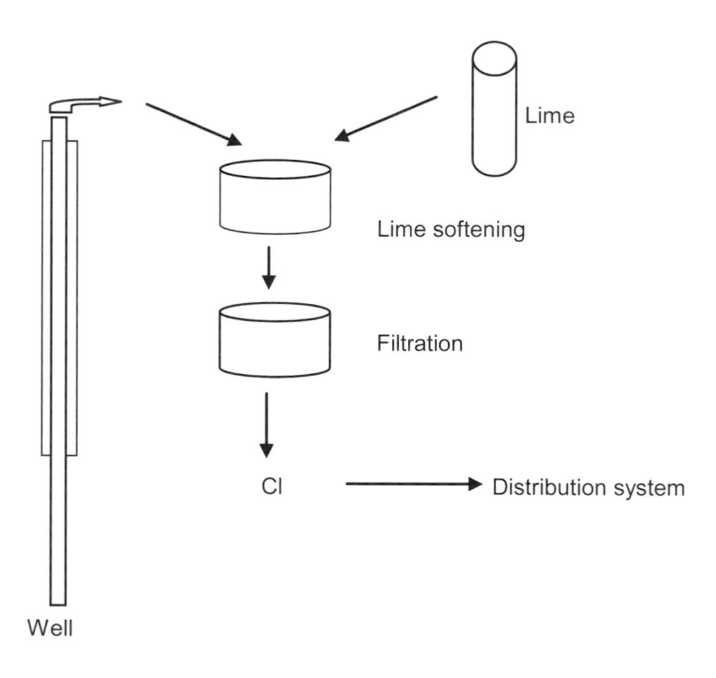

Figure 5-2 Average-quality groundwater treatment regime
Raw water has high hardness, limited metals, treatable microbiologicals, no volatile organic compounds/synthetic organic chemicals, and generally meets SDWA requirements.

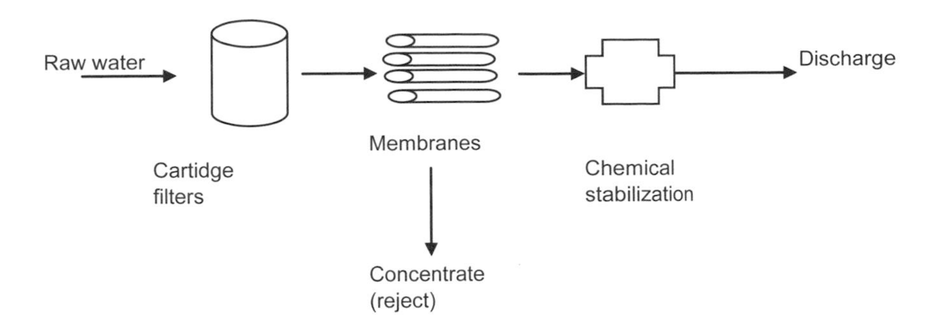

Figure 5-3 Membranes for dissolved-solids removal

(typically chlorides). These aquifers are termed brackish. Membranes are typically used to treat salty water (see Figure 5-3), although other desalination processes exist. Groundwater quality in many sedimentary basins (where the older and deeper sediments were deposited by oceans) can change very abruptly in mineral content. Poor-quality water can be drawn upward after production begins (upconing), even if a production well does not penetrate the formation. The wells can also induce saltwater intrusion into freshwater aquifers by drawing the saltwater front horizontally toward the wells.

An aquifer contaminated with volatile organic compounds (VOCs) will require air-stripping. Such aquifers usually have minerals and other problems and are considered low quality (see Figure 5-4). Federal or state regulations require treatment for public health reasons. Many water systems also install special treatment to meet public demand for better water quality. The following paragraphs outline the more common constituents to be removed and the methods for doing so.

Hardness

Hardness, the result of dissolved calcium and magnesium, is the most common water quality problem in wells. The hardness of water is expressed in parts per million (ppm) or milligrams per liter (mg/L). Hardness ranges from 40 mg/L to hundreds of parts per million. Most water systems try to distribute water with hardness in the range of 75 to 150 ppm, which is considered moderately hard. Water that is too soft will create excessive bubbles when soap is used and excessive water use may result. Hard water has just the opposite effect—very hard water inhibits lathering by soap and causes a soap curd that precipitates in clothes and on plumbing fixtures. Hard water will form a buildup of scale in water distribution and hot water piping and boilers, which must be removed to avoid plugging. Lime softening is the standard process for removing hardness, although ion-exchange resins may be effective in some cases. Nanofiltration membranes are the newest technology designed to lower hardness. Where water is naturally very hard and softening is not provided by the water system, customers often install softening units on their water service.

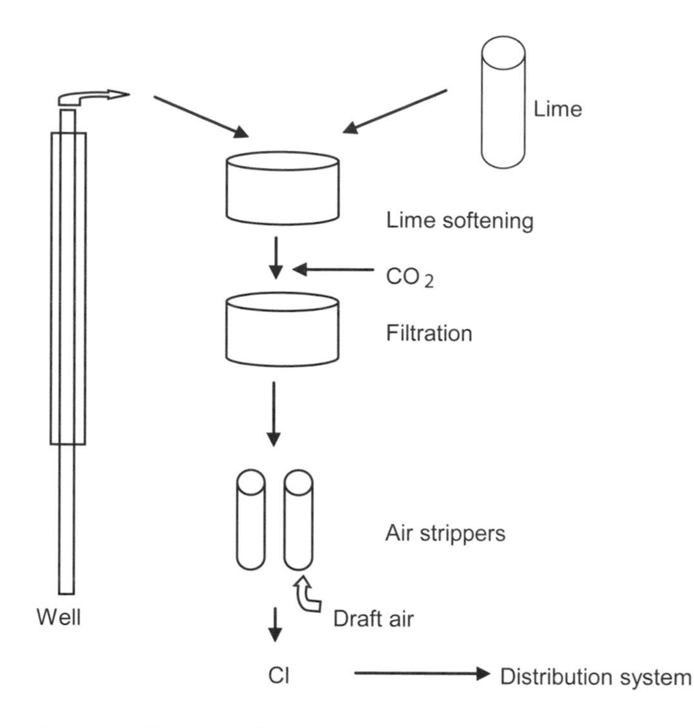

Figure 5-4 Low-quality groundwater treatment regime

Raw water has high hardness and metals, microbiologicals, volatile organic compounds (VOCs)/synthetic organic compounds (SOCs), sand/turbidity, and difficulty meeting SDWA requirements.

Iron and Manganese

Iron and manganese in high concentrations are objectionable in a public water supply because they cause stains in plumbing fixtures and laundry. They may also cause objectionable tastes and odors and difficulties in some manufacturing processes. Pyrite is a crystallized form of iron that may be an indicator of high levels of arsenic.

Because of the wide distribution of iron in nature, the presence of iron is common in groundwater. The iron is solubilized by groundwater and when pumped from a well is still in the soluble state. When exposed to oxygen, the iron is quickly converted to the oxidized form, similar to rust. Oxidized iron in water has a yellow-brown color. Manganese causes a similar brown color when oxidized unless sulfur is present, in which case it will be black. Softening can provide some help, but typically ion exchange, aeration, or some combination will be required.

Microbiological Activity

The presence of organics, iron, and manganese in water also makes a water system vulnerable to growth of bacteria. Bacteria were discussed in chapter 4. Microbiological accumulations/biofilms pose two significant concerns: the potential to harbor pathogens and damage to steel in the treatment and distribution systems. Bacteria can also damage equipment, cause taste and odor problems, and result in red or black water. High doses of disinfectant and a good flushing program can control bacterial populations in the distribution system.

Fluoride

Fluoride compounds are present to some extent in most groundwater. Although a low level of fluoride compounds is considered beneficial to protect teeth from cavities, levels in excess of 2 mg/L may cause discoloring or deformities in teeth. Water systems with more than 2 mg/L in finished water must periodically notify the public of the potential problems. Fluoride concentrations in excess of 4 mg/L are considered dangerous to public health. Public water systems having excessive fluoride levels may reduce the concentration to an acceptable level by blending water from high- and low-fluoride sources, by removal treatment, or by changing to another low-fluoride water source.

Radionuclides

Radionuclides occur naturally in groundwater in some parts of the United States. Federal maximum contaminant levels (MCLs) have been established for the allowable concentrations of radionuclides in drinking water to protect public health. A water system having wells with excessive levels of radium or uranium should investigate changing water sources or blending water with high levels of radionuclides with waters with low levels of radionuclides to maintain the level in water served to the public below the MCL. If no other source is available, removal of radium and uranium can be accomplished by ion exchange, but disposal of the ion exchange waste products is more challenging due to the radioactivity of the waste.

Radon

Radon is a radioactive gas that is present in water from many areas. Ingestion of the gas at low concentrations is not considered harmful to health. The gas can be liberated by the spray action of showers and other appliances. A buildup of radon gas in buildings can create a potentially serious health danger if the gas is inhaled. Radon can be removed by passing water through an air-stripping tower or through a bed of granular activated carbon.

Nitrates

Nitrates may be present at excessive levels in shallow wells, especially in agricultural areas. High nitrate levels usually come from septic tanks, sludge, or fertilizers that have contaminated the groundwater. The prime adverse health effect of high nitrate levels is that they can cause methemoglobinemia, or blue-baby syndrome, if the water is fed to young babies. If a well is found to have a high level of nitrate, removal is possible, but it is usually best to seek another water source.

Arsenic

Arsenic is a common metal that is both carcinogenic and acutely toxic. Arsenic concentrations vary considerably across North America. Arsenic is a groundwater issue associated with the dissolution of rocks, particularly those with high levels of pyrite. Federal regulations establish maximum arsenic levels permitted in drinking water supplies. Treatment for arsenic can be expensive. Ion exchange is effective with arsenic.

Methane

Methane can occur in groundwater and is dangerous because an accumulation of the gas can, under some conditions, cause an explosion. Aeration is the normal method to remove methane and other volatile organics from water supplies, but the gas must be carefully vented and adequate precautions taken to prevent explosions.

Salts

Deeper aquifers, and those located near the ocean, may have considerable amounts of dissolved solids and salts in the water. Salts are difficult to treat using conventional processes. Membranes are used to remove salt. In coastal areas, all wellfields should have ongoing saltwater intrusion monitoring programs to limit salt impacts. Many wellfields in coastal, southeast Florida have been lost due to reduced groundwater levels that encourage saltwater intrusion.

SURFACE WATER TREATMENT NEEDS

Just as groundwater may need treatment, surface water also needs treatment, the amount depending on the water quality. For high-quality surface waters, it may only be necessary to filter the water and disinfect it (see Figure 5-5). For more turbid water, coagulation processes and clarification will need to be employed (see Figure 5-6). More polluted waters, those perhaps having volatile compounds, may require air-stripping (see Figure 5-7). Note that with the exception of coagulation processes, many of these processes are also employed for groundwater, albeit the design considerations may be different.

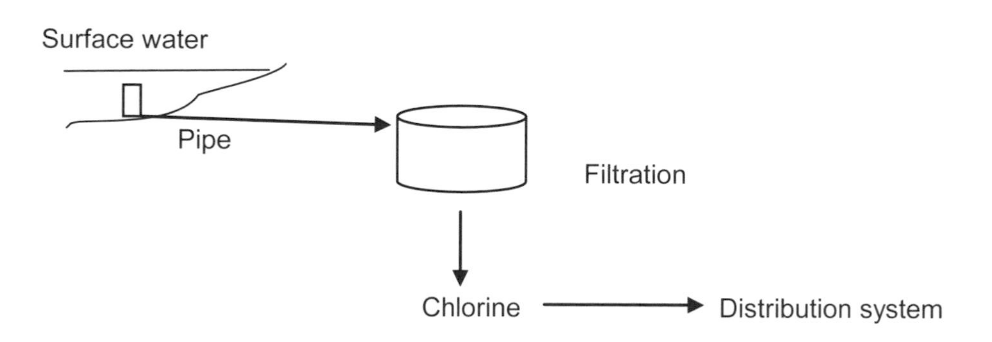

Figure 5-5 High-quality surface water treatment regime

Surface water has low hardness, low turbidity, low variability, no metals, significant source water protection efforts, cold temperatures, and meets SDWA standards.

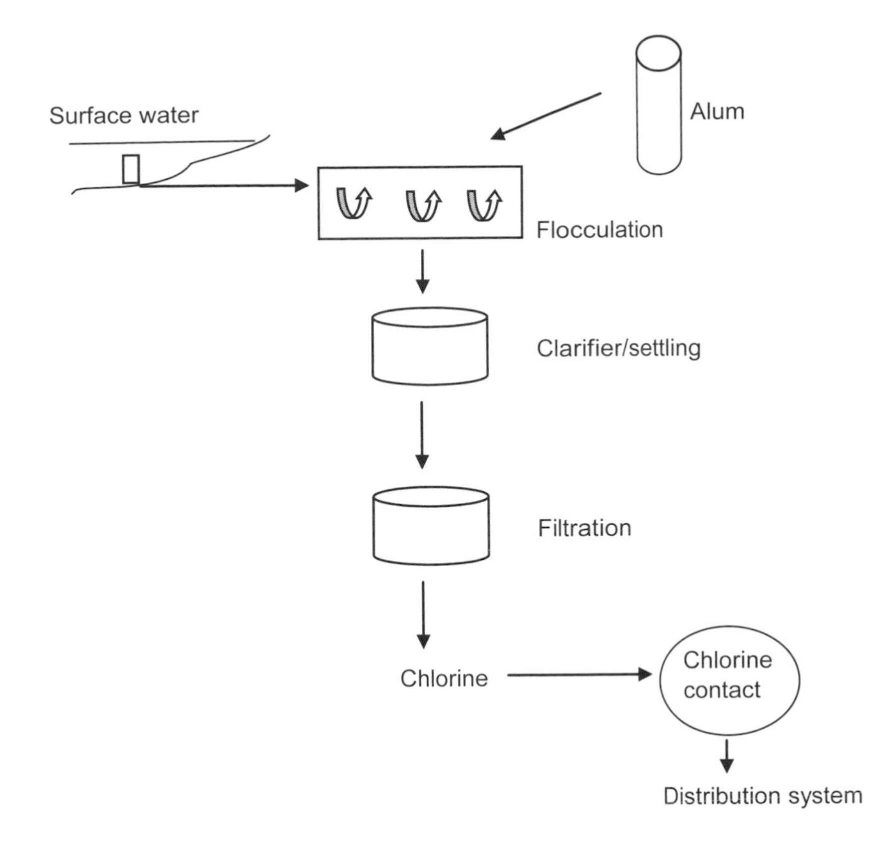

Figure 5-6 Medium-quality surface water treatment regime
Surface water has low hardness, generally low turbidity with seasonal variability, low metals, and limited source water protection, but otherwise meets SDWA standards.

The primary focus of surface water treatment is pathogen removal and/or inactivation. Pathogens can readily enter and survive in surface water. This means that turbidity in the water must be reduced. Turbidity is not removed by a chemical reaction like hardness is, but by clumping of particles to make them settle. Minimizing the risk to public health from harmful chemicals from industrial or agricultural activity is also important.

Turbidity

Turbidity is a measure of the cloudiness of water. The turbidity is typically organic, resulting from suspended soil, silt, and colloidal particles that leach out of the soil. The major problem in treating surface waters is that many have variable turbidity as a result of rainfall and snowmelt patterns. The small particles do not settle, but they frustrate disinfection efforts. Highly turbid water may have significant pathogens as a result of the suspended particles adsorbing the disinfectant. The particles also provide a means for pathogens to attach, which interferes with the ability of the treatment plant to disinfect the water adequately. During the spring or after major rainfall, turbidity may be very high, which may prevent effective disinfection of the water. From an aesthetic viewpoint, most

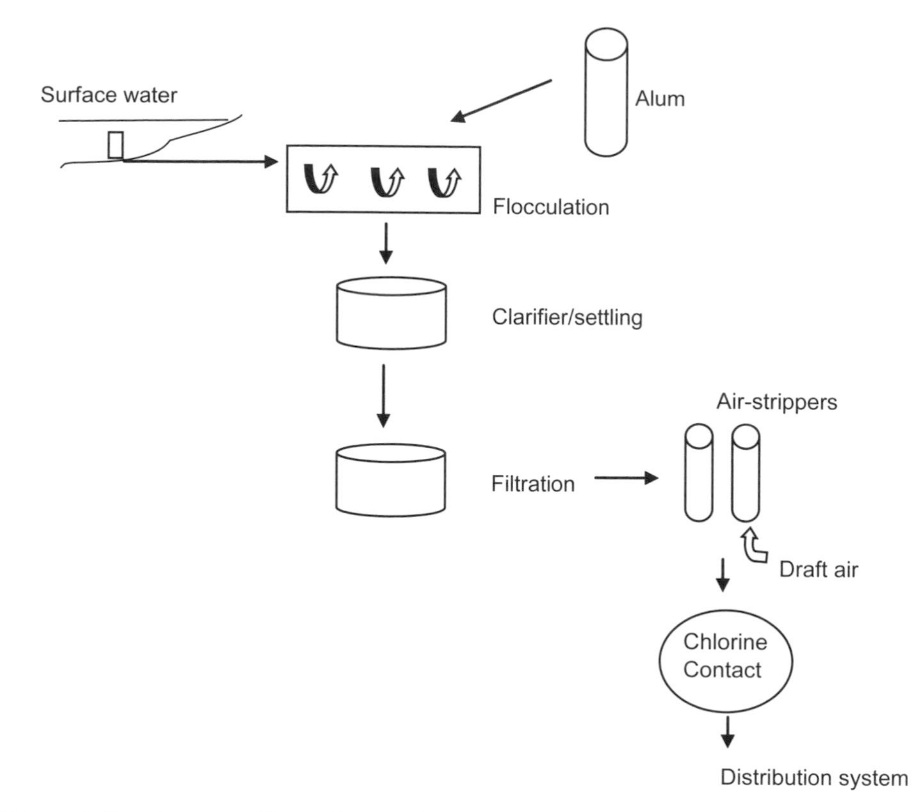

Figure 5-7 Low-quality surface water treatment regime
Surface water has high turbidity, metals, industrial waste contamination, and microbiological contamination, and does not meet SDWA standards.

customers would find visible turbidity in their water unacceptable, even if they were assured that it was safe to drink. For this reason, water quality regulations require surface water systems to meet specific limits on turbidity of water delivered to the distribution system. The addition of a coagulant like alum, ferric chloride, ferrous sulfate, and polymers is designed to encourage the particles to clump together (flocculate) and settle. Filtration also removes small particles and bacteria, which is why all surface water systems are required to have filtration as a part of the treatment process.

Tastes and Odors

Surface waters often absorb objectionable tastes and odors from algae and other decaying matter in the water supply. Algae concentrations indicate low oxygen in the water. Taste and odor problems of this type are most likely to occur in the summer. The problems are usually seasonal and short term and generally not harmful to human health, but the aesthetic problem with drinking water that tastes or smells bad is of concern and will cause complaints from customers. The goal of aesthetic standards is to provide water that will be perceived well so that people do not shift to waters of lesser quality but better appearance.

Tastes and odors can also be caused by industrial chemicals. Although clean water regulations require industries to limit or eliminate discharges to waterways that may create a public health or ecosystem concern, it is possible to have lake or river contamination by an accidental leak or spill, a broken pipeline, an overturned tank truck, or chemicals dumped into a sewage system and not removed by treatment. Some industrial chemicals, such as phenol, are accentuated by chlorine and are offensive to consumers.

Reservoirs and lakes can usually be treated with chemicals to inhibit algae growth. Keeping oxygen in the water and preventing exchanges of water due to temperature changes can be reduced by mechanical mixing. Systems using river water often construct enough water storage so that they can stop drawing raw water for the period when poor-quality water is passing the intake. Activated carbon and ozone can be used to resolve taste and odor problems that cannot be eliminated at the source.

Chemical Contaminants

Chemical contaminants considered harmful to human health are occasionally present in surface water sources. Some water systems close their intakes when chemical contaminants are present. Most chemical contaminants are caused by accidental releases or by runoff. Farm and lawn runoff can be serious where nitrates and pesticides run off into raw water supplies of any type. Nitrate levels occasionally exceed federal and state standards at times of the year when there is heavy runoff from agricultural lands. Endocrine-disrupting pharmaceuticals are an emerging area of concern.

Continual testing of surface waters will help operators choose the appropriate methods for nitrate treatment. Many small utilities may not test the raw water sufficiently. Operators should be trained on specific tests to detect the presence of all chemicals that could contaminate a surface water source. Source water protection efforts and monitoring of activities in the watershed or wellhead protection zone will provide the water system with a better understanding of potential contaminants.

TREATMENT METHODS

Treatment methods can be split into physical and chemical treatment processes, although in some instances both are required. Certain methods apply more to groundwater or surface water, while some apply to all sources. For instance, the most common chemical treatment methods for groundwater are lime softening, along with specific iron and manganese removal, carbon treatment, and ion exchange. For surface waters, alum and slow mixing (flocculation) are used to remove turbidity. In both cases the intent is to use the chemicals to precipitate the contaminants in the water.

Other chemical methods are used to achieve results such as disinfection or specific-ion removal. All waters require chemical disinfection, typically with chlorine, as a posttreatment process. Iron and manganese require chemicals to remove the ions. Other chemical processes are specific to the needs of the community.

Figure 5-8 Lime silo, flash mixer, and clarification zone

Physical treatment methods include clarification and sedimentation, filtration, membrane processes, and, to an extent, ion exchange. Surface waters require sedimentation and filtration in most cases. Groundwater may require no physical processes. The following pages outline the treatment methods most commonly used in water treatment plants across North America.

Chemical Lime Softening

The concept behind lime softening is to precipitate compounds that cause hardness by creating a chemical reaction with lime and the hardness particles. Lime is stored in dry form in a lime silo (see Figure 5-8). Lime is mixed with raw water to form floc (large particles), which settles quickly, thereby removing calcium and magnesium hardness along with iron and other metals, in a clarification zone (see Figures 5-8 and 5-9). Soda ash may be used in conjunction with lime to improve the efficiency of magnesium removal, but it should be used with caution since the concentration of sodium may become an issue. One of the problems with the chemical softening process is that it produces a by-product—lime sludge—that may prove difficult to dispose of after treatment because it dewaters (dries) poorly.

Figure 5-9 **Lime softening reactor**

Conventional (Alum) Treatment (Flocculation)

Conventional treatment is the term generally used for the combination of chemical addition, flocculation, sedimentation, and rapid sand filtration used by the vast majority of surface water treatment plants in the United States. Surface water is usually treated with alum addition (slow mixing and coagulation to remove suspended particles, as opposed to dissolved solids). The process is similar to lime softening but uses different chemicals and is intended to remove different materials. Surface water supplies suffer from suspended solids and turbidity in the water. Suspended solids only settle when large in size. Coagulation with alum is intended to cause the suspended material to coagulate into larger particles (called floc), which then settle in the clarifiers. Figure 5-10 illustrates the process, and the similarities are obvious. However, the flocculation basins are often separate from the clarifiers, and the flocculation basins are typically larger than those of lime-softening systems because the agitation action must be much slower to prevent the floc from breaking apart and the initial suspended-particle condition from persisting.

The clarifiers or sedimentation basins are designed to allow particles to settle. Detention time is long. Water is removed only at the top of the clarifier (see Figure 5-11). Circular clarifiers allow for an exponential decrease in particle velocity, which can make them more efficient. Clarifiers require a significant amount of space. Water from near the top of the clarifier is then directed to the filters. The sludge that has accumulated at the bottom of the clarifier must periodically be removed and disposed of. Drying and landfill disposal are common practices.

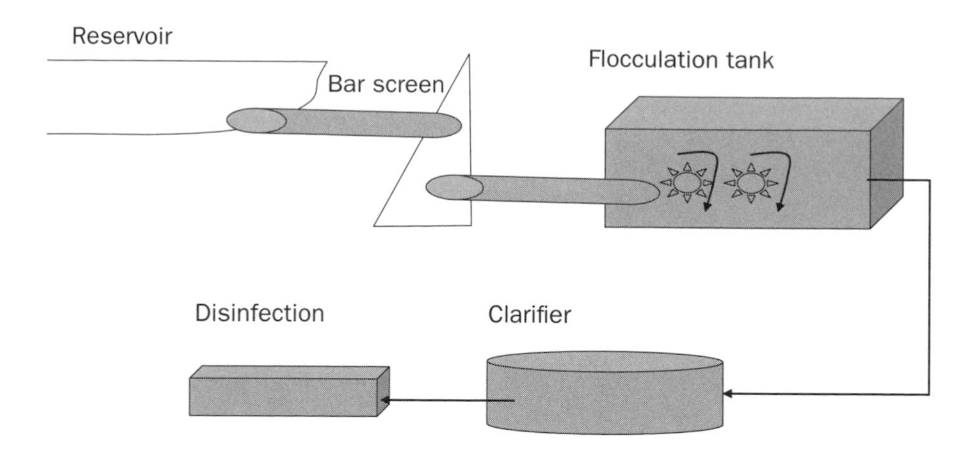

Figure 5-10 Conventional treatment

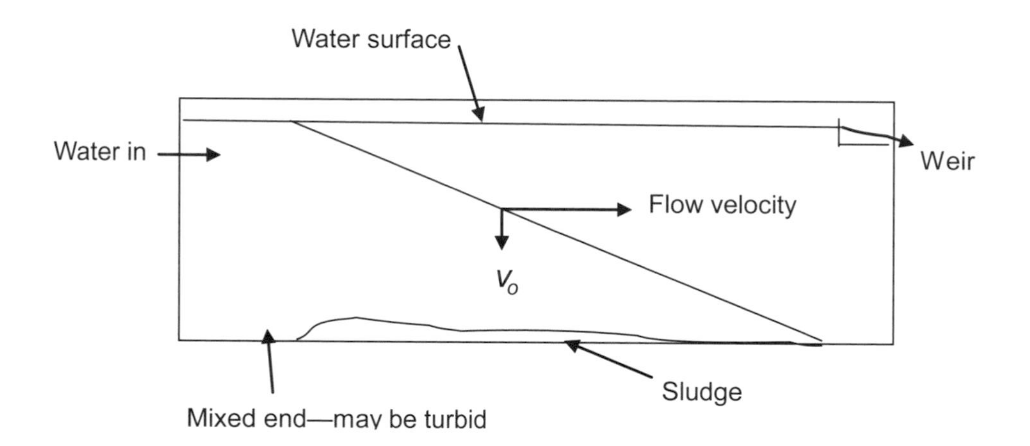

Figure 5-11 Clarification

NOTE: V_o is the settling velocity of the particle designed to be removed in the clarifier.

Unlike groundwater sources, the variability of surface waters requires much more extensive monitoring of the raw water and ongoing modifications to the treatment process to provide adequate treatment at all times, as alum addition and other chemicals must continually be adjusted.

Filters

Filters are used to remove particulates from the water in groundwater treatment systems where other treatment precedes it (such as lime softening). Filtration is needed in these circumstances because chemical addition creates a precipitate that is only partially removed in the reactors. Fine particles will not settle, meaning filtration must be employed to complete removal. Filters can be either pressure or gravity systems with fast or slow feed rates. All include a media that

Figure 5-12 Gravity sand filters

Figure 5-13 Backwash troughs in gravity sand filter

may include diatomaceous earth (DE), garnet, sand, or layers of sand and anthracite. Figures 5-12 and 5-13 show a pair of gravity rapid sand filters. Water is loaded from the top and allowed to filter though the sand. Figure 5-13 shows the backwash troughs. Periodically, the filter will start to plug. The filter then must have water flushed through it backward. When this happens, the sand bed expands and the dirt, silt, and colloids on the surface are flushed into the troughs and removed. The filter is then ready to operate as a clean filter again. Some systems use pressure filters that operate in a similar manner (see Figure 5-14). They are totally enclosed.

Figure 5-14 Pressure filters

Rapid Sand Filters. Rapid sand filters are constructed using sand of special grading and grain size so that the water will pass through rapidly. Some newer designs of rapid sand filters use two or more layers of sand and other media of varying specific gravities to form filters that can be operated at high flow rates and have increased capacity to hold particles before plugging. Treatment plants are usually furnished with two or more filters so that water continues to be processed through the plant while a filter is backwashed.

The quantity of silt and turbidity in most surface water sources is so great that it would plug a rapid sand filter in a very short time. A sedimentation process is therefore usually used to settle out as much of the suspended matter as possible before the water is filtered.

Slow Sand Filters. Slow sand filters were the first type of filter used by public water systems for treating water. The raw water is directed to large beds of sand where most of the suspended matter is removed as the water seeps through to perforated piping below. When the bed becomes clogged, the top 1 to 2 inches (25 mm to 50 mm) of sand is removed by hand or with mechanical equipment and disposed of. After a portion of the original bed has been removed, the sand is replaced. Two of the prime restrictions on the use of slow sand filters are that the raw water quality must be relatively good and a large amount of space must be available for the treatment system because of the large size of the filters.

Pressure Filters. Pressure filters are enclosed vessels that push water through the sand medium under limited pressure. Figure 5-14 shows a pressure filter. Pressure filters may have higher ratings than gravity filters. However, backwashing is more frequent. Large pressure filters are usually designed to automatically backflush.

DE Filtration. DE filtration uses a relatively small filter container having a porous membrane, or septum, through which the water must pass. Diatomaceous earth is a mined product consisting of the crushed shells of microscopic sea animals. It is usually necessary to also add a continuous body feed of diatomite during filtration in order to maintain porosity of the filter cake and extend the filter run. DE filters are widely used for swimming pool filtration. For drinking water treatment, DE filters are considered appropriate only for direct filtration of surface water with low turbidity and low bacterial levels.

Direct Filtration. Direct filtration, in its simplest form, consists only of adding a coagulant chemical, mixing, and passing the water through rapid sand filters, with no sedimentation step. This process can only be used on surface water systems where there is relatively little suspended matter in the source water. Such filters require frequent backwashing, especially if turbidity increases even to a minor extent.

Avoiding Filtration. Federal regulations allow some public water systems that use surface water sources to avoid filtration treatment if they can demonstrate that their water source is consistently very clean, not turbid, and has minimal potential for pathogens. The principal requirement for avoiding filtration is that the watershed for the water source must be protected from contamination caused by human activities, including recreation. Disinfection treatment of the water must be continuously carried out to meet exacting requirements, and the water system must meet detailed monitoring and reporting requirements. A water system that has been granted permission to operate without filtration but later fails any of the mandatory requirements may be directed by the state or the US Environmental Protection Agency to install filtration within 18 months. Some states have a general requirement that all public water systems using surface water must use filtration.

Ion Exchange

The ion-exchange process is also used for softening. The concept is to pass water through large tanks containing an ion-exchange media—a larger version of home water softeners. Typically the discharge from the softening system is blended with the hard, raw water to produce finished water of medium hardness. Salt is used to regenerate the exchange media during a backwash cycle. Two advantages of ion-exchange softeners are that they produce a minimal amount of by-product and they can be installed at individual well locations in small systems.

Air-Stripping/Aeration

Many industrial chemicals found as contaminants in groundwater are volatile chemicals. Volatile organics, as well as radon, readily escape from the water if it is aerated. The older methods of aeration included cascading water down trays of coke and pumping the water into the air over large tanks, as shown in Figure 5-15. More complete air-stripping is now provided by aeration towers. The towers are tall tanks with the water to be treated flowing in at the top and a large volume of air blown in at the bottom (see Figure 5-16). The tanks are filled with

Figure 5-15 Well and aeration system on top of storage tank

Figure 5-16 Aeration tower

plastic balls or other shapes that continually redirect the water as it flows downward through the tank (see Figure 5-17). By adjusting the air-to-water ratio, very high aeration efficiency can be obtained. An aeration tower does not take very much space, does not need to be enclosed, and has a relatively low operating cost. The primary problems in locating a tower adjacent to a well are that it may not be visually acceptable to nearby residents and that being open may expose water in the well to bacterial contamination.

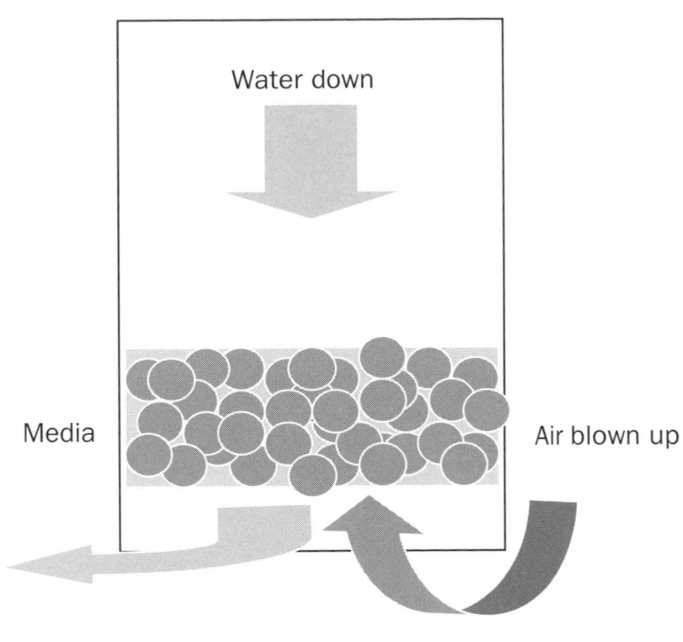

Figure 5-17 Aeration concept

Membranes

Figure 5-3 (p. 39) shows a typical process flow diagram for membrane processes. Membranes are physical, like filters, but remove particles down to molecular size. Because the particles removed are so small, membrane plants usually have two filtering processes. Cartridge filtration is essential for removal of suspended particulates larger than 5 μm from the raw water. Once the raw water is chemically conditioned and suspended solids have been removed by the cartridge filters, it is delivered to the feed pumps as *feedwater*. A dedicated feed pump supplying each skid increases the feedwater pressure prior to applying the feedwater to the membranes themselves. Figure 5-18 shows a typical membrane skid. The membranes are housed in the horizontal tubes shown in Figure 5-18.

Figure 5-19 details the membrane unit and how it works. On one side is the clean water (permeate) while the other side retains the minerals (concentrate). Pressure pushes the raw water through the membrane's very tiny holes.

Table 5-1 outlines the types of membranes available. For water service, reverse osmosis (salt) and nanofiltration (softening) are the ones used. Typical skids accommodate two or three stages of membranes. Each stage recycles the concentrated water from the previous stage. The more stages, the higher the recovery rate. The process recovery rate for nanofiltration (membrane softening) is 85 to 92 percent, while brackish reverse osmosis (saltwater removal) systems can produce only a 50 to 75 percent recovery, as compared to minimal losses in more traditional treatment regimes. The reduced recovery in the saltwater processes increases the quantity of raw water required to produce the same amount of permeate from one process skid, while producing a larger waste stream of concentrate, meaning that the concentrate disposal requirements are greater

Figure 5-18 Membrane system (softening or salt removal)

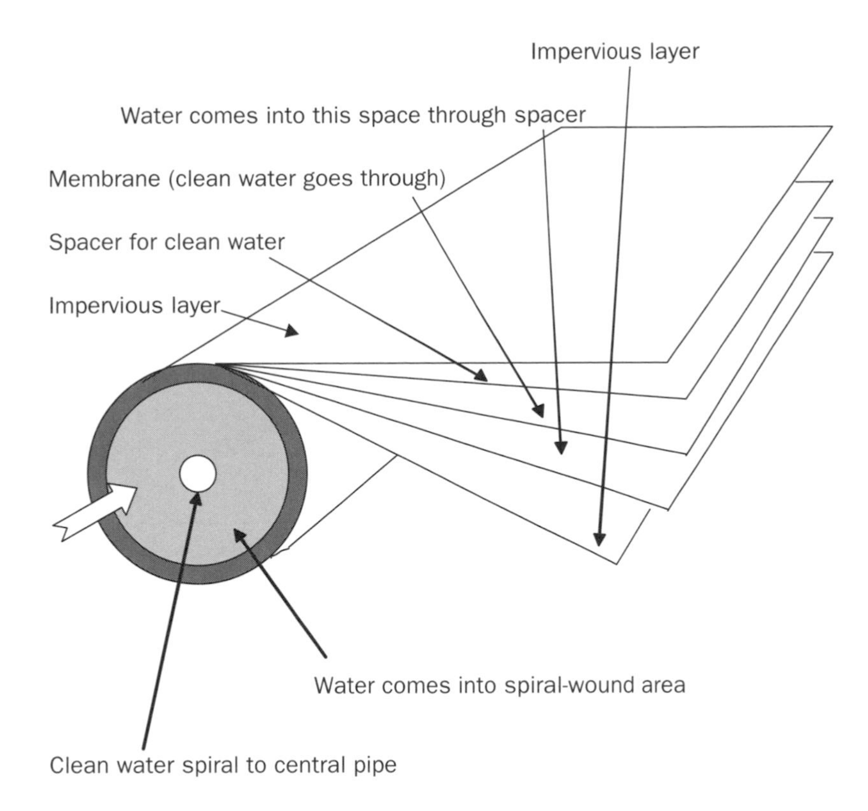

Figure 5-19 How spiral-wound membrane elements work

and more raw water is needed to serve the same number of people. Disposal of the concentrate is a major concern with all membrane systems as the ionic imbalance of the concentrate makes it acutely toxic to marine and freshwater organisms.

A membrane cleaning/flushing system is required, which consists of cleaning and flushing solution tanks, 5-µm cartridge filters, and cleaning pumps. The cleaning pumps are constructed to handle high- and low-pH cleaning chemicals. The cleaning system must be designed to accommodate future needs when the system is expanded. If a membrane system is shut down, a shutdown flush is required so that any raw water in the membrane elements is replaced with permeate water.

Iron and Manganese Control

If iron and/or manganese levels in groundwater are not too high, it is often possible to control discoloration by adding an oxidizing chemical such as chlorine or potassium permanganate. The iron and manganese precipitate that is formed can then be removed by settling, then passing the water through a filter. Another option is to add a sequestering agent to the water before it is disinfected or the water is exposed to air. A sequestering agent prevents the iron and manganese from oxidizing and causes them to remain in solution for a period of time. Water systems have varying degrees of success with sequestering agents.

Table 5-1 Types of membrane processes

Type of Membrane	Pore Size	Particles Removed	Pressure (psi)	Permeate (%)
Microfiltration	1–10 µm	0.5–1 log virus removal turbidity, cysts	5–30	95–98
Ultrafiltration	1–100 nm 100 K MW	4 virus removal macromolecules	7–60	80–95
Nanofiltration	100 MW	Softening	80–120	70–90
Reverse Osmosis	10 MW	Salt, ions	200–1,200	50–85

If a product is found that works, very little investment is required for equipment and cost for chemicals is relatively low. Aeration can also be used to remove iron and manganese by reaction with oxygen.

Carbon Treatment

When tastes, odors, or color in surface water cannot be adequately controlled by conventional chemical treatment or by proper selection of disinfectants, water quality can usually be improved by using activated carbon. It has been known since ancient times that passing water through a bed of charcoal removes tastes and odors. A benefit of activated carbon is that it adsorbs many harmful industrial and agricultural chemicals as well as trihalomethanes where present, adding a further treatment method to reduce contaminants to acceptable levels. Activated carbon comes in two forms: granular and powdered activated carbon.

Granular Activated Carbon. Granular activated carbon (GAC) is a source of carbon that has been prepared to make the surface of the particles very porous. The extremely small pores have the ability to trap many harmful industrial chemicals as well as pesticides and herbicides as the water passes through a bed of GAC. GAC treatment of groundwater has been used successfully in situations where a well has chemical contamination and no alternate source of water is available. Large tanks containing GAC can be quickly brought to a site and connected to a well supply to remove the contamination while further testing is done and a permanent solution to the problem is developed. In general, long-term use of GAC for contaminant removal is expensive. If a contaminant can be removed by aeration, the operating cost is much lower. If a contaminant can only be removed by GAC and there is no other source of water available, the process may have to be used continuously.

Surface water systems with continuing taste and odor problems or chemical contamination sometimes find it advantageous to use GAC as a coarse media in the filters, where it serves a dual function of both filtration and contaminant removal. Other surface systems pass the water through complete conventional treatment and then direct it through tanks filled with GAC before the water enters the distribution system. When GAC has become so plugged that it no longer functions, it must either be reactivated by a special heating process or disposed of and replaced with new material.

Powdered Activated Carbon. Powdered activated carbon (PAC) is similar to GAC except that it is available as a very fine powder that is fed into the treatment process as opposed to being in a filter bed. PAC will absorb most taste- and odor-causing substances in raw water. The carbon is usually fed near the beginning of the treatment process, and most of it settles out with the floc in the sedimentation basin.

POSTTREATMENT PROCESSES

The major posttreatment process is disinfection. Disinfection is the final protective barrier against pathogens in the water being delivered to the customer's tap. All raw water sources have the potential for contamination by pathogens. Surface water runoff contains pathogens for obvious reasons, but even properly constructed wells may still become contaminated by waterborne disease–causing organisms if the surface contamination can get into a well through rock fissures, unnoticed defects in the well seal or pump installation, rusting out of the well casing, or contamination during well maintenance or repair.

Pathogens are organisms commonly found in water and soil that may cause human sickness or death. Many can remain viable for a considerable period of time despite the efforts of treatment processes. Detecting pathogens on a real-time basis can be difficult, so an indicator test is used—the detection of the presence of fecal coliform bacteria. These organisms are used because they provide an indication of the presence of wastes from humans and warm-blooded animals. They are excreted from warm-blooded organisms in huge quantities. Tests for the presence of other specific microorganisms such as viruses and *Giardia* cysts are difficult to use, costly, and not as commercially available as fecal coliform tests, so they should be used in conjunction with other detection methods.

All public water systems are required periodically to perform coliform tests to provide assurance that the water has been adequately treated and the system has not become contaminated. However, coliform tests are not always the best indicator of pathogen safety because coliforms are relatively short lived and are easily killed by chlorination, while viruses and oocysts may not respond to chlorination at all.

Disinfection is a relatively inexpensive way of ensuring that a system regularly meets microbiological monitoring requirements and of minimizing the possibility of a waterborne disease outbreak. The requirement for maintaining a chlorine residual is to prevent water from becoming contaminated by microorganisms while in the distribution system.

Disinfection

Disinfection usually occurs as the last step in the treatment process. Of all the available disinfectants, chlorine is the only one that provides a substantial residual that continues to be active all the way to the customer's tap. For this reason, free chlorine is used as a disinfectant by most groundwater systems in the United States. Small systems sometimes feed chlorine as a solution, similar to household bleach. Larger systems obtain the chlorine as a gas in cylinders (see Figure 5-20) and add it to the water through a special device known as a chlorinator (see Figure 5-21). Chlorination equipment can be automated to work unattended, with proper safeguards to prevent freezing and vandalism.

Some chlorine may be added at the initial treatment stages for color control or to improve chemical treatment processes. One problem with the use of chlorine is its potential to form trihalomethanes—a group of chemicals that are formed by the reaction of chlorine with organics that may be present in the water. Trihalomethanes are cancer-causing chemicals, so the allowable concentration in drinking water is regulated. All water systems are required to have samples analyzed periodically for the total trihalomethane concentration. Trihalomethane formation may be limited through the use of ammonia to create chloramines.

Figure 5-20 Gas chlorine cylinders

Figure 5-21 Chlorine injector pumps

Disinfectants most commonly used for drinking water treatment are free chlorine, chloramine, chlorine dioxide, hypochlorite, ozone, ultraviolet (UV) light, and potassium permanganate. Each has advantages and disadvantages, such as the cost of feed equipment, operating costs, ease of use, disinfection potency, effects on tastes or odors in water, and the tendency to create disinfection by-product chemicals in the finished water. It should be noted that groundwater only requires secondary disinfection, which eliminates UV light, ozone, and potassium permanganate as disinfectants.

Free Chlorine. Free chlorine is the most widely used drinking water disinfectant in the United States. The principal advantages of chlorine are that it is a very strong disinfectant and a persisting residual of chlorine continues disinfecting as the water passes into the distribution system. Free chlorine is delivered as a liquid under pressure in cylinders. The liquid is evaporated in special equipment and injected into the stream of water as a gas in most cases. Chlorine is also moderately priced and relatively easy to use. The cylinders can pose a health risk if they leak, but special precautions are made by the chlorine producers to minimize risks. To wit, in the 100 years of use of chlorine gas at water plants in the United States, there has only been one operator death associated with the use of chlorine gas. Trihalomethanes are more likely to form when combined with free chlorine residuals.

Chloramines. Chloramines are a somewhat different form of chlorine that is formed when chlorine reacts with ammonia. Two benefits exist with using chloramines: they last longer than free chlorine, and they do not form trihalomethanes as readily as free chlorine, a benefit where water may have high organic loading. While in some cases ammonia is present in the raw water, most water systems purposely form chloramines by adding ammonia to the water before or after the chlorine feed. Chloramines may work better than free chlorine in controlling tastes and odors, but the disinfecting capability of chloramines is much weaker, and they are considered particularly weak in inactivating certain oocysts and viruses. Recent concerns about the formation of *N*-nitrosodimethylamine (NDMA) during chloramination are the source of continuing research on this disinfection method.

Chlorine Dioxide. Chlorine dioxide may be manufactured as needed in the treatment plant by the interaction of chlorine with other chemicals. It is a strong disinfectant and is used by some water systems because it is less likely to accentuate tastes and odors than free chlorine, and it reduces the formation of trihalomethanes.

Hypochlorite. Hypochlorite comes in two forms: liquid sodium hypochlorite (bleach) or calcium hypochlorite tablets (such as are used in swimming pools). Tablet forms of hypochlorite must be converted to liquid to use them effectively. Hypochlorite is commonly used to disinfect water lines and should be available on all water systems for this purpose. Small water plants can use hypochlorite in place of chlorine gas; it poses a lesser operating risk to the public while maintaining the public health at the tap. Bulk forms of hypochlorite are commercially available.

Ozone. Ozone is extremely reactive and can effectively eliminate tastes and odors in water when used as an initial disinfectant. It quickly dissipates, though, so there is no disinfectant residual, meaning that chlorine in some form must be added as well as the ozone. Ozone also creates by-products that have not been studied significantly. Because ozone must be generated as needed, the cost of equipment and operations (electricity, principally) is considerably higher than for chlorine. Ozone is widely used in Europe, principally due to waste loading in raw water supplies. Ozone has not been used by many water systems in the United States for disinfection because chlorine must be added to obtain a residual. Ozone does provide some benefits in reducing color, although ozone creates some by-products that are the subject of current research.

UV Light. UV light is occasionally used for disinfecting drinking water for very small applications. UV technology is increasing in use, but since there is no residual, chlorine must be added. UV appears to be useful for inactivation of bacteria and viruses, but the quality of the raw water must be high to achieve good disinfection, meaning filters and membranes may be needed.

Potassium Permanganate

Potassium permanganate is used by many surface water systems as an oxidant and in groundwater systems to reduce color. It is not widely used for disinfection. It has the advantages of being easy to feed and handle, reducing the formation of trihalomethanes, and acting to remove tastes and odors by oxidation in the source water. It does not form a usable residual, so systems using permanganate must also feed chlorine to provide the required disinfectant residual in the distribution system.

Fluoridation

Studies over the years have shown that fluoride strengthens children's teeth and reduces tooth decay. While various methods of providing a continuing dose to children have been tried, the most successful method is to add fluoride to drinking water in the range of 0.9 to 1.2 ppm. For this reason, most states/provinces now have requirements for public water systems to maintain the optimum level of fluoride in water furnished to the public. Many groundwater sources contain some naturally occurring fluoride, so these systems only need to add the amount necessary to bring the fluoride concentration up to the state-mandated level. Most surface water sources have little or no naturally occurring fluoride, so it is necessary to add the desired amount during the treatment process. Water systems may apply fluoride to the water by using a concentrated acid or by dissolving one of several types of fluoride chemicals that are available. The costs of fluoridation feed equipment and chemicals are relatively small compared to overall system operating costs.

Corrosion Sequestering and Control

Corrosion control may be required to control the potential for aggressive water to dissolve lead, copper, and other compounds in the piping system. Federal regulations enacted in 1991 in the United States require special monitoring for lead and copper in customers' drinking water. Where excessive levels are found, special corrosion control treatment must be provided. Understanding the source of the problem helps water systems add appropriate chemicals to resolve it.

Groundwaters tend to be basic (pH greater than 7.0) and deposit calcium carbonate on pipes. Lime softening and recarbonation for pH adjustments can disrupt the natural tendency of the water to deposit carbonates (too much deposition will reduce effective pipe diameter and reduce flows). Likewise, some surface waters are relatively acidic and may cause iron and steel pipe and tanks to disintegrate and lead in old plumbing and water service pipes to dissolve. In either case, the tendency of the water to cause corrosion can be corrected by the addition of chemicals to adjust the water. The Langelier index is typically used for this purpose. A neutral Langelier index is optimal. A wide range of products are available to control corrosion, but the water system should test several that appear to be good choices and use one that optimizes corrosion control. The utility should be careful in these tests, because choosing the wrong sequestrant may cause failure in lead and copper samples.

This page intentionally blank.

Chapter 6

WASTEWATER TREATMENT

Unlike water treatment, wastewater treatment starts out as neither a chemical nor a physical process—it is a wholly biological phenomenon and therefore has a different set of challenges and a capacity for disruption under extreme weather and a variety of operating conditions. The basic treatment units include aeration basins, clarifiers, and disinfection facilities. The disinfection and clarification processes are substantially the same as for water systems outlined in chapter 5, but totally different concepts are included in the aeration basins.

There are a variety of different biologically oriented treatment processes in wastewater plants, but the differences are primarily oriented toward the length of time the waste stays in the aeration basin. Otherwise the processes generally operate under the same set of variables: food (the waste coming into the plant); microorganisms that consume the waste and convert it into new cells, carbon dioxide, and water; and oxygen, which is required for the organisms to respirate. Organically based wastes come into the treatment facility with trace amounts of minerals, metals, and other contaminants. The aeration basin contains bacteria that have been "trained" to use these organic wastes as "food." The amount of "food" for the bacteria is measured as the carbonaceous biological oxygen demand (CBOD). For most applications, a five-day test is used (termed BOD_5 or $CBOD_5$). To survive, the bacteria require oxygen, so air is pumped into the aeration basin to improve respiration. The amount of air and food will determine an optimal microorganism population. Maintaining this food-to-microorganism ratio is important for healthy bacterial populations; the goal is to keep the bacteria healthy but close to starvation so that they consume the waste and reproduce efficiently. In the aeration basin, there will be active areas of bacteria growth and areas where the organics have been removed from the water.

The wastewater generally flows into the aeration basin on one end and flows out the other. On the exiting end, a *clear* zone is present—this is the *treated* water, which is decanted to the clarifiers, where the remaining materials are allowed to settle to the bottom before being recycled back to the aeration basin with the incoming organics. People often wonder what settles. As the bacteria age and die, they descend to the bottom of the aeration basin or clarifier as sludge. When sludge is removed from the facility, it is actually minerals and dead bacteria, viruses, rotifers, and so on—not the organic material that comes into the wastewater treatment plant.

Additional equipment often includes mechanically cleaned bar screens, grit tanks and filters for the removal of solids that cannot be consumed by the bacteria, nitrification and denitrification facilities to reduce nutrients (primarily ammonia and nitrate), effluent pumps, and disposal systems. The effluent quality required dictates the equipment needed. The following paragraphs describe the process requirements at various levels of treatment. Disposal of the treated wastewater and the potential receiving water's use indicate the extent to which treatment is required. Regulatory agencies in some areas have made nutrient removal a priority in fresh waters and estuary systems, which significantly increases the amount of treatment required for those systems.

DEGREES OF TREATMENT

Primary Treatment

Primary treatment is defined as the use of the treatment system to accomplish the removal of a portion of the suspended solids and organic matter prior to discharge into the receiving water (Figure 6-1). No biological processes are assumed to occur. Typically, the treatment consists of settling basins (or primary clarifiers) and macroscale screening (i.e., bar racks or screens). These processes remove only the largest constituents (and those most likely to clog pumps and pipes). Thus the effluent will have a high concentration of CBOD and organics (typically over 40 percent of the incoming amount). Primary clarifiers are designed to remove 50 to 70 percent of the suspended solids and 25 to 40 percent of the CBOD. They typically precede biological processes and can be used for flow equalization in secondary treatment facilities. These clarifiers have a detention time of only 10 to 30 minutes, hence the low removal rates compared to secondary treatment. In the United States and Canada, most primary facilities have been phased out and replaced with secondary treatment systems. Primary treatment plants are not appropriate for developed areas.

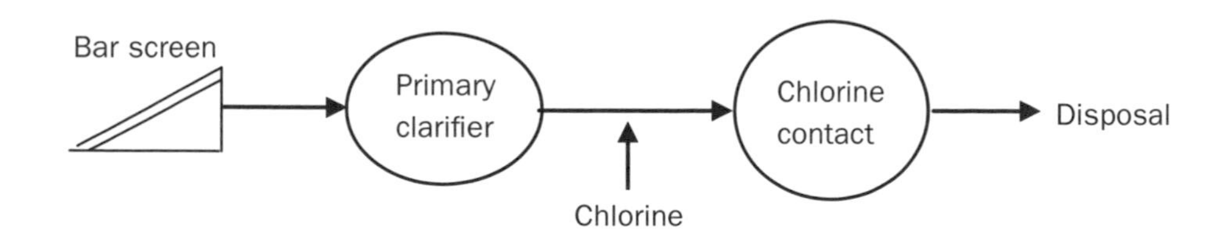

Figure 6-1 Primary wastewater treatment

Secondary Treatment

Secondary treatment principally describes the removal of biodegradable organics ($CBOD_5$) and total suspended solids (TSS). Biological processes include activated sludge, fixed-film reactors, and extended aeration systems. Typically, a secondary treatment facility has a bar screen, and it may have primary clarifiers ahead of the biological treatment process (Figure 6-2). The input of oxygen by mechanical means (aeration) is a fixture of secondary treatment. Aeration may come in the form of air bubbles injected through diffusers near the bottom of the aeration basin or mixers on the surface of the aeration basin. All employ secondary clarifiers after the biological process. Aeration basins and clarifiers together define secondary treatment plants. Disinfection (typically via chlorination) is generally employed. Most facilities in the United States and Canada are secondary treatment systems.

Secondary plants are designed to achieve an effluent prior to discharge containing not more than 30 mg/L $CBOD_5$ and 30 mg/L TSS, or 85 percent removal of these pollutants from the wastewater influent, whichever is more stringent. The requirement is 20 mg/L $CBOD_5$ for injection wells. Appropriate disinfection to minimize the impact of potentially pathogenic organisms and bacteria in the receiving water and pH control of the effluent are normally required. Downstream users, and their proximity to the discharge, may dictate more stringent treatment. Coastal waters have more stringent effluent limits (see the Advanced Wastewater Treatment section).

Reclaimed Water (Advanced Secondary) Treatment

Advanced secondary treatment is reused, reclaimed, or recycled-quality water (Figure 6-3). It requires the employment of all secondary processes, plus filtration and high-level disinfection, which is defined as a chlorine residual over 1.0 mg/L after a given period of time (typically 15 to 30 minutes). Typically, the filtration step uses gravity sand/anthracite filters. Land application

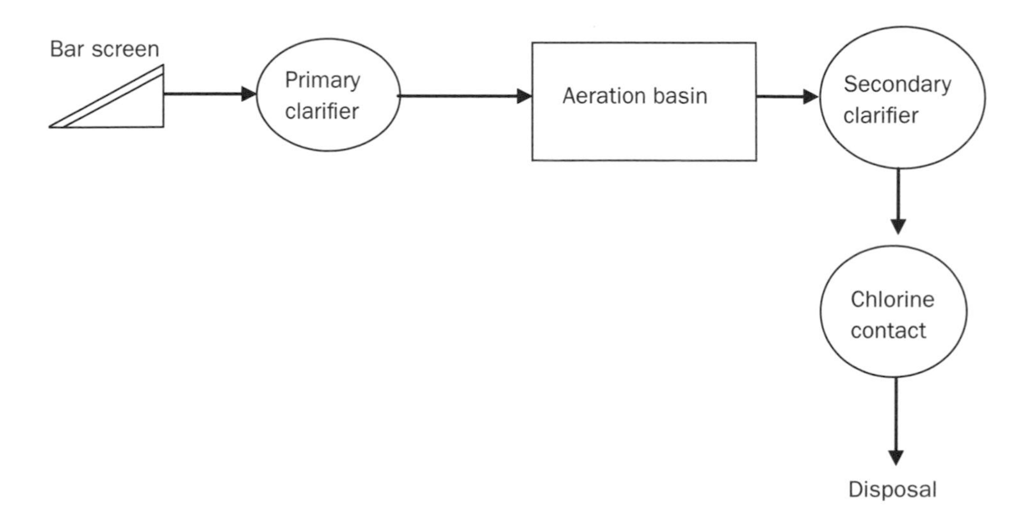

Figure 6-2 Secondary wastewater treatment

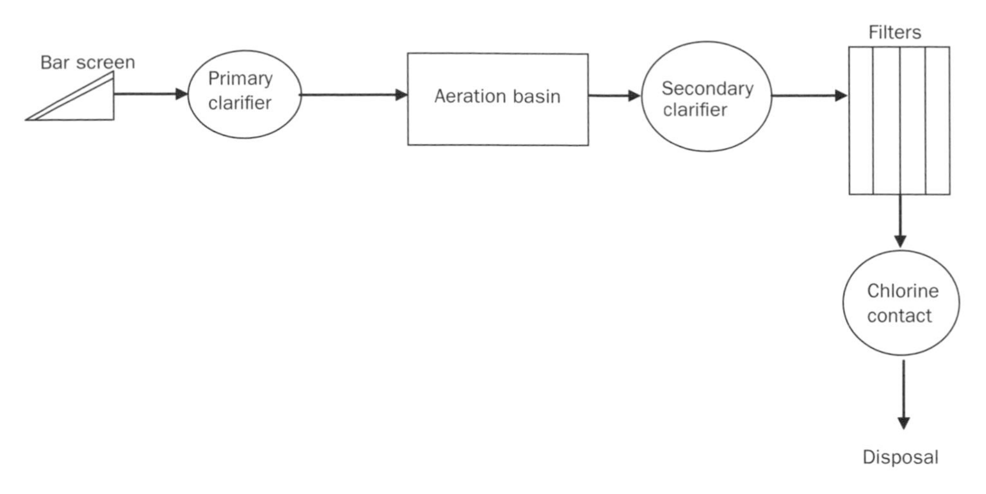

Figure 6-3 Advanced secondary wastewater treatment

or groundwater discharge systems (excluding underground injection) are designed to achieve an effluent quality (after disinfection) containing not more than 5 mg/L TSS. Advanced secondary treatment is often confused with tertiary or advanced wastewater treatment. However, the latter assumes nutrient removal, which does not occur with filtration. For reuse-quality disposal, the reuse recipient (often a golf course, park, or residence) generally prefers nutrients in the water.

Principal (More Advanced Secondary) Treatment

Principal treatment is a term, coined in Florida for reclaimed water projects, to be applied to groundwater recharge and indirect potable reuse projects. Principal treatment standards require all advanced secondary treatment requirements, but total nitrogen is limited to 10 mg/L as a maximum annual average. This is accomplished by means of chemical feed facilities, which include coagulants, coagulant aids, polyelectrolytes, or an anoxic zone in the aeration basin. Chemical feed facilities may be idle (on standby) if the water quality limitations are being achieved without chemical addition. Figure 6-4 shows the process.

Advanced Wastewater Treatment

The term *advanced wastewater treatment* (AWT), as generally used, includes the treatment necessary to raise wastewater quality beyond that produced by secondary treatment (including reduction of nutrients, toxicity, suspended solids, and organics). Typically, AWT includes secondary treatment plus nutrient removal (nitrification, denitrification, and phosphorus removal) and may not contain more, on an annual average basis, than the following concentrations (see Figure 6-5):

- $CBOD_5$ 5 mg/L
- TSS 5 mg/L
- Total nitrogen (as N) 3 mg/L
- Total phosphorus (as P) 1 mg/L

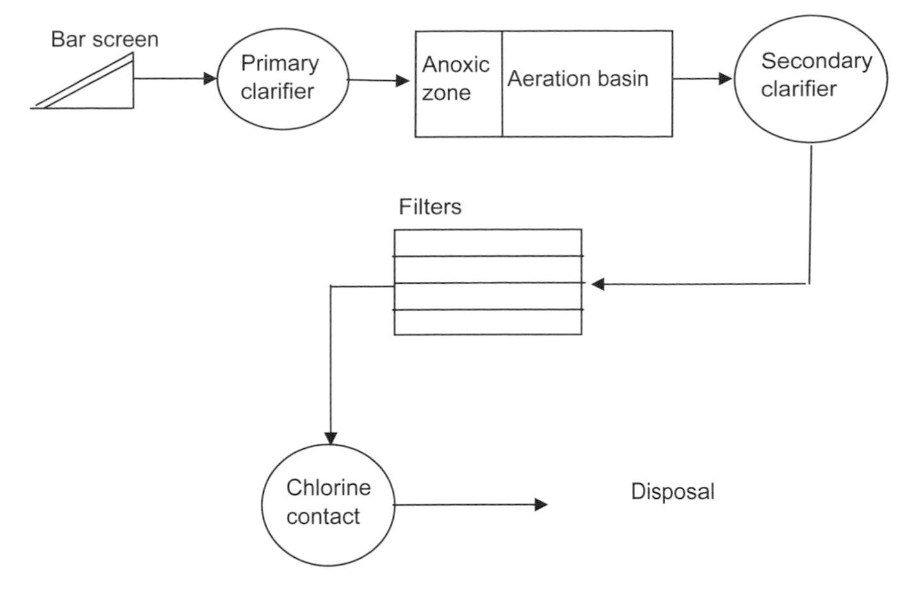

Figure 6-4 More advanced secondary wastewater treatment

TSS removal can be accomplished via the filtration step in advanced secondary treatment. Nitrification/denitrification is commonly accomplished with methanol and bioreactors or rotating biological disks. Phosphorus is commonly removed by the addition of alum coagulants (which increases sludge volume significantly). There is typically a requirement to dechlorinate the effluent to minimize toxicity effects in the receiving water. This level of treatment is commonly applied to nutrient-sensitive surface waters, coastal estuaries, groundwater recharge programs, and indirect potable reuse projects.

Full Treatment and Disinfection

This concept is employed in groundwater recharge programs in California and Florida. *Full treatment* systems include all the treatment steps contemplated under AWT plus reverse osmosis and/or activated carbon for removal of the remaining organics to reduce total organic carbon (TOC), total organic halogen (TOX), and some pathogen removal (see Figure 6-6). Dechlorination is often required. The requirements are as follows (Chapter 62-610, Florida Administrative Code; FDEP 2006):

- The parameters listed as primary drinking water standards are applied as maximum single-sample permit limits, except for asbestos.
- The primary drinking water standards for bacteriological parameters are applied via the disinfection standard.
- The primary drinking water standard for sodium is applied as a maximum annual average permit limitation.
- Except for pH, the parameters listed as secondary drinking water standards are applied as maximum annual average permit limits.
- All pH observations must fall within the pH range established in the secondary drinking water standards.

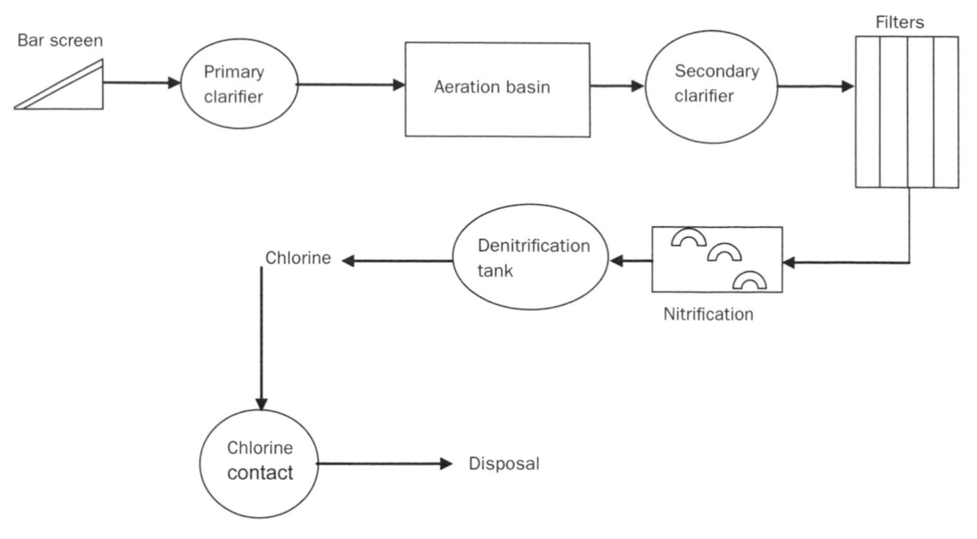

Figure 6-5 Advanced wastewater treatment

- Additional reductions are required of pollutants, which otherwise would be discharged in quantities that would reasonably be anticipated to pose risk to public health because of acute or chronic toxicity.
- Total organic carbon cannot exceed 3.0 mg/L as the monthly average limitation; no single sample can exceed 5.0 mg/L.
- Total organic halogen cannot exceed 0.2 mg/L as the monthly average limitation; no single sample can exceed 0.3 mg/L.
- The treatment processes must include processes that serve as multiple barriers for control of organic compounds and pathogens.
- Treatment and disinfection requirements are additive to other effluent or reclaimed-water limitations.

This is obviously very expensive treatment to pursue and therefore would be an option only in areas with very limited water resources. Water Factory 21 in Orange County, Calif., is an example water system that has used these processes.

TREATMENT PROCESSES

The following paragraphs outline specific biological treatment processes. Activated sludge and contact stabilization are the most commonly constructed biological treatment facilities at this point in time, although many old trickling filter plants are still in existence. Fixed media and nitrification processes are generally used today for more advanced treatment processes.

Activated Sludge

Activated sludge organisms are designed to effectively metabolize the organic food in a wastewater plant. Activated sludge is commonly used for large treatment plants. Large amounts of air are pumped beneath the surface through

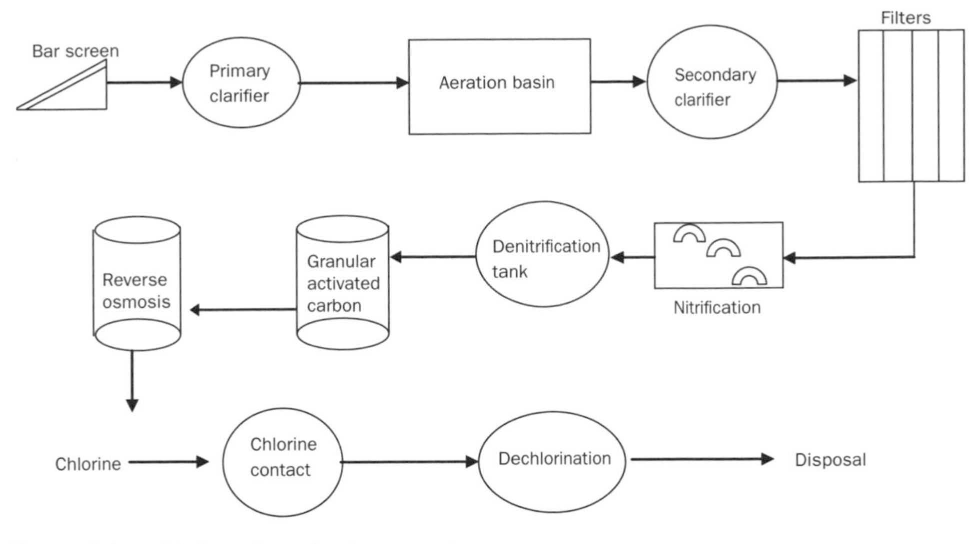

Figure 6-6 Full wastewater treatment

diffusers to provide oxygen to the organisms. Pure oxygen is used to reduce air demands in some large treatment plants to reduce basin sizes. The microorganisms generally react quickly to the loading; thus contact time averages four to six hours. The amount of sludge created is significant, and it is recirculated to the head of the plant on a significant scale (30 to 50 percent of the total flow as returned activated sludge is not uncommon). Activated sludge is an efficient treatment process, but it requires operator attention. Plant upsets, where the microorganisms are disrupted, washed out, or killed, are perennial potential problems resulting from excessive inflow from rain or high levels of toxins from industrial users. Monitoring of the treatment process and a degree of operator sophistication are required for activated sludge facilities.

Oxidation Ditches (Extended Aeration)

Oxidation ditches, or extended aeration systems, are easy to operate and are forgiving of variability in the collection system and weather patterns. The intent of an oxidation ditch is to provide more than 18 hours of retention time for the microorganisms to attack the food sources. Air is generally introduced by mechanical means—brushes or disks. Oxidation ditches are efficient for small utility systems where plenty of land is available, but over 10 million gallons per day (mgd), the cost of land and tankage makes oxidation ditches not cost-effective. Figure 6-7 is a typical oxidation ditch facility. It is a popular choice for utilities whose staff has limited experience with wastewater operations.

Contact Stabilization

There are thousands of ring-steel contact stabilization plants, such as the one shown in Figure 6-8, throughout the world. They are efficient systems that include digesters, reaeration, and disinfection systems in a neat, circular structure. Plants commonly come in sizes up to 2 mgd. Multiple contact stabilization

Figure 6-7 Oxidation ditch

systems can be placed on one site. However, equalization of flows becomes a problem when multiple parallel systems exist on one site. The small footprints and efficient operations are benefits of contact plants. The system is not as forgiving as an oxidation ditch, but it remains relatively easy to operate. High-quality effluent can be produced in contact stabilization plants that are operated well under capacity.

Fixed-Film Media

Trickling filters and rotating biological contact (RBC) units are common fixed-film systems. The concept with fixed-film systems is to minimize moving parts. The bacteria, instead of being in a solution, are attached to the media—usually redwood, plastic, or stone. Air is forced up from the bottom as water cascades down in a trickling filter (similar to an air-stripper as outlined in chapter 5). The media is slowly rotated in an RBC, alternately submerging and revealing the bacteria. The processes are efficient for low flows, but sloughing of the microbiological colonies can be a problem. Many RBCs have been converted to use in nitrification systems. The systems are not as efficient or controllable as activated sludge, contact stabilization, or extended aeration facilities.

Nitrification/Denitrification Systems

Nitrogen is the next nutrient removed after carbon has been substantially removed from the wastewater. Nitrification is a process whereby microorganisms are encouraged to convert ammonia to nitrate and nitrite. Denitrification is

Figure 6-8 Contact stabilization plant

a process whereby additional carbon is added to force the microorganisms to convert the nitrate to nitrogen gas, thereby eliminating it from the effluent. RBCs and methanol, along with tertiary clarifiers, are used to accomplish these goals. Removal of nitrogen is obligatory in many coastal and estuarine areas at this time. These systems can be difficult to control where the collection system is variable. They require staff experience to operate.

DISPOSAL OF EFFLUENT

Disposal of wastewater effluent drives the treatment process required. Surface waters have been, and remain, the most popular choices for disposal of wastewater effluent. However, the impacts downstream of the wastewater discharge, especially where there are stressed ecosystems, water utilities using the same stream for their water supplies, or economic issues that require higher-quality water, may require significantly more treatment than the wastewater is currently receiving. The downstream issues continue to evolve as scientists find more impacts. Endocrine-disruptor effects on fish, amphibians, and birds in water bodies is the latest issue that is being investigated.

Ocean outfalls are used in 18 states, but they are limited in most of those states. Estuarine impacts are documented, and as a result, ammonia reductions are required. Open-ocean outfall discharges do not yield the same evidence, but they continue to be studied. Obviously, ocean outfalls are not available for many states. Likewise deep injection wells are used in a number of states. Limitations due to geological formations exist in most states, thereby preventing the use of injection wells for waste disposal.

The most popular alternative in water resource circles is reuse. However, the need for irrigation varies dramatically from day to day and month to month. The highest-usage times do not often occur with the highest supply periods (which tend to be during the rainy season due to pipe infiltration in the collection system). There are also ongoing investigations into the impact of reuse on water supplies (viruses, endocrine disruptors, and so on) that as yet have not fully reached maturity. As a result, while reuse has potential benefits (especially to industrial clients and power companies for cooling), it still needs a backup alternative for times when the demands for reuse are low. Other options include agricultural irrigation, groundwater augmentation, salinity barriers, and a host of other uncommon or site-specific options. All have similar issues to reuse water.

PRETREATMENT PROGRAMS

All utilities are required to meet stringent limitations on the quality of treated wastewater that is disposed of. The intent of such programs is to help prevent business enterprises from discharging substances into the wastewater treatment system that cannot be treated, will cause the treatment organisms to die, will cause the effluent limitations to be exceeded, or will otherwise be detrimental to or upset the plant process or operations. Substances regulated include such things as oil, gas, metals, cleaning fluids, paint, and industrial process water. Pretreatment is mandated under the Clean Water Act in the United States and may be a condition of the utility's discharge permit if perceived by the regulatory agency to be an issue. Size of system, types of customers, and effect on receiving waters will determine this requirement. The impact of pretreatment programs can affect the ability of the community to attract businesses, so careful consideration of effluent limits and treatment efficiency is required.

Chapter 7

WATER PLANT OPERATIONS

Several tasks are included in plant operations beyond just the operation of treatment plant equipment. Maintenance of the equipment, water quality monitoring, record keeping for permit compliance, and waste disposal are key elements of utility operations. The following paragraphs outline these responsibilities.

PLANT OPERATORS

The treatment plant operator has the following general responsibilities:

- Check, adjust, and operate equipment such as pumps, meters, in-line water quality analyzers, and electrical systems
- Determine chemical dosages and keep chemical feed equipment charged with chemicals, adjusted, and operating properly
- Perform routine maintenance and condition checks of equipment and make minor repairs
- Order and maintain a stock of parts, chemicals, and supplies
- Maintain operating records and submit operating reports to the system owner or responsible person and to the state
- Perform tests and special analyses required for proper operational control
- Collect and submit samples required by the state at the proper time
- Keep informed of federal and state regulations affecting the water system
- Recommend to superiors any major repairs, replacements, or improvements to the plant that should be made

The plant operator of a small utility should not spend all of his or her time just operating the water production or sewage treatment equipment. His or her duties should include maintenance, painting, maintenance logs, work orders, chemical analyses, and coordinating with other sections of the utility groups, such as water distribution, sewage collection, and engineering. Most states have requirements as to the amount of time the operators must spend at the treatment plant each day performing these tasks. In addition to performing or directing water production (or sewage treatment), operation, and maintenance, the operators may also direct operation of the water distribution system, operate the sewage treatment plant and sanitary sewer system, or work at other municipal functions.

In larger and more complex treatment systems, operators spend more time performing the duties of operating and maintaining the production facilities. Surface water treatment plants usually require closer monitoring and have more operations that must be performed manually. States generally require that an operator be on duty while a surface water plant is in operation unless special monitoring equipment is installed.

PLANT MAINTENANCE

It is particularly important that water treatment equipment be properly maintained to minimize failure and protect the system from threats. A well, pump, or piece of treatment equipment that fails because of improper maintenance can be very costly and disruptive to customers. For this reason, maintenance of important pieces of equipment should be performed regularly as recommended by the manufacturer. This should be accompanied by frequent inspection and testing to anticipate failure or degenerating performance. As an example, periodic review of well records can help to anticipate repairs and schedule maintenance work.

Major maintenance of equipment, whether done by the utility or by contract, is best done at a time of year when water or sewer use is low. This enables the system to meet normal operating demands while the equipment under repair is out of service. Preventive maintenance is a positive program to pursue as it generally lengthens the amount of time that mechanical equipment operates and minimizes breakdowns. Preventive maintenance also may provide operators with an indication of failure in the near future, allowing them to schedule maintenance prior to the failure.

WATER QUALITY MONITORING AND REPORTING

As noted in prior chapters, one of the most important duties in operating a water system is the continual analysis of the water provided to the customer. These water quality analyses include fecal coliforms to determine if enough disinfection has occurred, primary and secondary water quality standards, trihalomethanes, and special parameters (often performed triennially) for volatile and synthetic organics. Most of these analyses are discussed in other parts of this handbook; further information can be gained from local public health agencies. Table 7-1 shows water quality analyses for utilities in southeast Florida.

It should be noted that no utility ever delivers water that is free of all other constituents—that would be distilled water. Distilled water is inherently unstable because it is undersaturated. As a result, distilled water is highly corrosive; it literally will dissolve copper, lead, and cement linings in pipes. Moderately hard water that tends to deposit calcium carbonate is preferable. South Florida water systems provide such water.

Some states may have state field staff collect chemical samples in conjunction with a sanitary survey or a special visit to the system. Other states furnish the operator with sample containers and instructions, indicating where and when samples must be collected and shipped. Still other states furnish the operator with only instructions on the samples that must be collected and a list of certified laboratories where samples can be sent for analysis. Instructions must be carefully followed to ensure that samples are collected at the proper location and time and that containers are properly filled and shipped. In cold weather, precautions must be taken to prevent samples from freezing.

Special care should be employed when sampling to avoid contamination of the samples, a common problem. Persons sampling the system should be trained on the proper techniques. Splitting samples in two, and retaining one of the splits, is good practice in the event of problems with the analysis. Split samples should be retained using the guidelines of *Standard Methods* (Eaton et al. 2005). Care should also be exercised when choosing a lab for the analysis—cheaper is rarely better.

For most analyses, the laboratory must analyze a sample within a set number of hours or days following collection. If a sample arrives at the laboratory past the required holding time, new samples must be collected. In some areas of the country, it may be necessary to hand deliver or ship samples by special parcel service to ensure that they arrive safely and on time.

When the laboratory has completed the analyses, it will send a report to the operator. If the laboratory has not furnished a copy directly to the state, the operator must send a copy to the state within a required number of days. The report will usually provide a comparison between the analysis results and established maximum contaminant levels (MCLs) for each parameter. The system's consultant or state staff can provide further explanation of laboratory results.

Under state and federal requirements, failure to perform required monitoring is usually cause for the system to be directed to perform public notification. This is usually both costly and embarrassing to the operator and public officials. To avoid having to perform public notification, care should be taken to strictly follow state-directed requirements. Water systems that provide treatment must also provide on-site analyses for various parameters, both to satisfy state requirements and to provide proper operation and control of the process. Surface water systems must make periodic analyses for turbidity level, and all systems that add disinfectant must analyze disinfectant residual concentrations as required by the state. Systems that provide fluoridation must also periodically analyze for the fluoride level in water supplied to customers.

Many larger water systems operate their own laboratories to provide some or all of the required analyses. The system's laboratory must be operated by a qualified technician and certified by the state for performing each type of analysis.

Table 7-1 Example of comparative water quality analyses in southeast Florida

Constituent	Units	Standard	Davie Beach	Dania	Broward County A	Broward County 2	Miramar East LS	Miramar Central M	Tamarac	Pembroke Pines	Hallandale Beach	Sunrise Springtre
		STD	DAV	DAN	BC 1	BC 2	MIR E	MIR C	TAM	PPM	HAL	SUN S
Cyanide Total	mg/L	ND	ND	ND	ND	ND	ND	ND	ND	ND	ND	ND
Gross Alpha	mg/L	15	ND	ND			0.3	ND	7.6		0.4	
Thms	ug/L	100	82	57							12	13
Aluminum	mg/L	0.2		0.07			0.057	0.01				
pH	mg/L	8.5	8.93	9.58	7.6	8.6	8.8	8.6	9.2	7.99	9.4	8.9
Radium 226	mg/L						0.1	0.1				
Radium 228	mg/L						ND	1.2				
MBAS	mg/L	0.5	0.03	0.054	0.1	0.057	0.044	0.1	0		0.038	ND
Arsenic	mg/L	0.05	ND	0.015	0.0007	ND	ND	ND	ND	ND	ND	ND
Barium	mg/L	2	ND	ND	0.003	0.006		ND	ND	ND	ND	ND
Beryllium	mg/L	0.004	ND	ND	ND	ND	ND	ND	ND	ND	ND	ND
Cadmium	mg/L	0.005	ND	ND	ND	ND	ND	ND	ND	ND	ND	ND
Chromium	mg/L	0.01	ND	ND	ND	ND	ND	ND	ND	ND	ND	ND
Copper	mg/L	1	ND	ND	ND	ND	ND	ND	ND	ND	ND	ND
Iron	mg/L	0.3	ND	ND	0.017	0.011	0.037	ND	ND	ND	ND	0.04
Lead	mg/L	0.015	ND	ND	ND	ND	ND	ND	ND	ND	ND	ND
Manganese	mg/L	0.05	ND	ND	ND	ND	ND	ND	ND	ND	ND	ND
Nickel	mg/L	0.1	0.006	ND	ND	ND	ND	ND	ND	ND	ND	ND
Selenium	mg/L	0.05	ND	ND	ND	ND	ND	ND	ND	ND	ND	ND
Silver	mg/L	0.1	ND	ND	ND	ND	ND	ND	ND	ND	ND	ND
Zinc	mg/L	5	ND	ND	0.003	ND	0.01	0.098	ND	ND	ND	ND
Mercury	mg/L	0.002	ND	ND	ND	ND	ND	ND	ND	ND	ND	ND
Antimony	mg/L	0.006	ND	ND	ND	ND	ND	ND	ND	ND	ND	ND
Fluoride	mg/L	4	0.85	0.8	0.8	0.85	0.85	0.84	0.81	1.08	0.7	1.12
Nitrate	mg/L	10	0.18	0.38	0.02	ND	ND	ND	ND	0.15	0.18	0.21
Nitrite	mg/L	1	ND	ND	ND	ND	ND	ND	ND	0.03	ND	0.03
Sodium	mg/L	160	34.6	22	34.1	19.4	14	15	23	17	15.2	40.1
Thallium	mg/L	0.002	ND	ND	ND	ND	ND	ND	ND	ND	ND	ND
Sulfate	mg/L	250	14.8	23.9	5.6	21.7	17	3.8	14.6	8.4	12	15.3
Color		15	12.5	13.3	4	4	11	2	3..5	ND	10	10
Odor		3	ND	ND	1	1	1.4	1.4	ND	ND	ND	ND
cis 1 2 Dichloroethene	ug/L	0	ND	ND	ND	ND	ND	ND	8.8	ND	ND	ND
4 Bromo-fluorbenzene	ug/L	0	ND	ND	ND	ND	ND	ND	ND	ND	ND	ND
Chlorides	mg/L	250	65	70	76	51	35	31	42.4	43.8	26.7	52.7
TDS	mg/L	500	206	230	208	186	340	54	176	163	122	256

TDS = Total dissolved solids
ND = Not detected
NA = Information not available
Thms = Trihalomethanes

Ft. Laud.	Deerfield	Pompano Beach	Margate	Plantation	Plantation	Lauderhill	Hollywood	North Spr. Im. Dist	Coral Sp Im Dist	Sunrise Pk. City	North Lauderdale	Cooper City
FTL	DFB	PB	MAR	PLN E	PLN W	LHL	HWD	NSID	CSID	SUN P	NL	CC
ND	ND	ND	ND	N	ND	ND	ND	ND	ND	ND	0.2	ND
		6	ND				3					
37	29.6					30	8.6			35		
0.033		0.009					ND				0.47	0.01
7.29	9.38	8.6	9	7.3	6.4	8	8.3		8.4			
		0.1	0.2				0.6					
		0.3	ND				0.5					
0.031	0.04		ND									0.014
ND	ND	0.001	ND	ND	ND	ND	ND	ND	ND	ND	ND	ND
0.19	ND	0.008	ND	ND	0.008	ND	ND	ND	ND	ND	0.005	ND
ND	ND	ND	ND	ND	ND	ND	ND	ND	ND	ND	ND	ND
ND	ND	ND	ND	ND	ND	ND	ND	ND	ND	ND	ND	ND
ND	ND	0.0008	ND	ND	ND	ND	ND	ND	ND	ND	ND	ND
ND	ND	ND	ND	ND	ND	ND	0.003	0.007	ND	ND	ND	ND
	ND	0.07	ND	ND	0.12	ND	ND	ND	ND	ND	ND	ND
ND	ND	0.0004	ND	ND	0.0015	ND	ND	ND	ND	ND	ND	ND
ND	ND	0.01	ND	ND	0.0012	ND	ND	ND	ND	ND	0.0007	ND
ND	ND	ND	ND	ND	ND	ND	ND	ND	ND	ND	ND	ND
ND	ND	0.002	ND	ND	ND	ND	ND	ND	ND	ND	ND	ND
ND	ND	0.3	ND	ND	ND	ND	ND	ND	ND	ND	ND	ND
ND	ND	0.01	ND	0.17		0.016	ND	ND	ND	ND	ND	ND
ND	ND		ND	ND	ND	ND	ND	ND	ND	ND	ND	ND
ND	ND	0.0005	ND	ND	ND	ND	ND	ND	ND	ND	ND	ND
0.93	0.93	0.93	0.6	0.99	0.99	0.64	0.77	0.92	0.27	0.93	0.5	0.88
0.3	ND	0.8	ND	0.05	ND	ND	0.02	ND	ND	0.12	ND	ND
ND	ND	0.06	ND	ND	ND	ND	ND	ND	ND	ND	ND	ND
20.6	21	71.35	26	15	14	17.1	19	31.1	35	49	33.2	47
ND	ND	0.0005	ND	ND	ND	ND	ND	ND	ND	ND	ND	ND
ND	16.6	13.9	17	2.6	2.91	12.6	ND	29.3	24.4	2.26	17.2	18.2
3.3	n/a	10	20	5	5	10	3	20	15	20	20	1.9
ND	ND	ND	1	ND	1	1	1	ND	ND	ND	2	ND
ND	ND	ND	4	0.36	ND	0.6	ND	ND	ND	ND	ND	ND
ND	ND	ND	101	6	ND	ND	ND	ND	ND	ND	ND	ND
61.5	43.1	46.8	54	1.06	10.7	18.6	38	63.4	64.5	67.8	51.6	17.5
240	107	208	200	27	47	132	180	312	296	259	208	120

Reporting to State Agencies

Each state has special report forms to be used by community public water systems to record operating and monitoring information. Different forms are available for various types of systems or treatment. The system operator is usually required to report operating data on every day that the system is in operation and submit reports to the state monthly. A copy of all reports should be kept for the water system files.

WATER TREATMENT WASTE DISPOSAL

Until the 1960s, it was common practice for water systems to discharge their treatment wastes back into the lake or river being used as a water source or into any other available watercourse. Plants employing conventional treatment almost always discharged their filter backwash water and sedimentation basin sludge without treatment. Obviously, the Clean Water Act prohibits this kind of activity, because discharging turbid or pathogen-laden water into a surface water body, especially one that could be used later for a raw water supply, could prove problematic. Most water treatment wastes can cause discoloration of the water and may have an effect on biological life in a lake or river. The following paragraphs outline several waste products that operators need to deal with.

Backwash Water

When filters plug, water is washed backward to clean the surface of the filter. Figure 5-13 showed the backwash troughs that remove the dirty water from the filter. Most treatment plants now recycle the backwash water to the influent of the plant to blend it with incoming water. This protocol conserves water and has little effect on the treatment process, as the suspended matter in the backwash water is eventually removed by the sedimentation process. There is little need to find other methods of disposal, although sanitary sewers may be available.

Sludge

Disposal of sludge is problematic. Use of sanitary sewers has been pursued in the past, but the potential for line plugging remains a concern. Most water plant sludges are harder to dewater than their wastewater counterpart, because the water plant sludge-producing processes tend to hold water. Alum sludge is particularly hard to dewater. Most large facilities dewater their own sludge and dispose of it in landfills, while small systems may still send their nonlimed sludges to the sewer system. Land application or burial of the sludge is sometimes used for disposal where there is no other practical alternative available.

Lime sludge is not sent to the sewer system, because it can be disruptive to the waste treatment process and lime will build up and block the sewers. Lime sludge is usually dewatered and buried at landfills or other sites. Unfortunately it is rarely "dry"; it is more like the consistency of cheesecake, which is problematic for landfill operators because it is easier to compact dry materials.

State authorities should be consulted on allowable methods of disposal, permits required, and monitoring requirements before a new disposal method is used for all sludges. Environmental laws are strict, and violators may be subject to heavy fines and possible liability for an environmental problem that they have created.

Membrane Concentrate

With the advent of membrane treatment systems, operators have to deal with the concentrate. Unfortunately, the concentrated products are ionically imbalanced and are therefore acutely toxic to aquatic organisms. The US Environmental Protection Agency (USEPA) regulates membrane water treatment plant concentrate like a hazardous material, which severely restricts the options for disposal despite its apparent passivity.

The difficulty in disposing of membrane concentrate is one reason the amount of waste should be minimized. Multistage membrane systems can lower reject rates substantially—three-stage nanofiltration can achieve over 90 percent recovery. Having lesser amounts of waste improves the likelihood that a lower-cost solution can be found for its disposal. At present, most membrane systems must use injection wells, wastewater plants, or very large bodies of water for dilution (such as thermal discharges from power plants), which is expensive. If the water system is large and the wastewater system is receptive and has capacity, this is a good option. However, concentrate may disrupt wastewater organisms, so an industrial pretreatment permit should be employed. The water system may need to chemically treat the concentrate to prevent damage to the wastewater system.

Ion-Exchange Backwash

Ion-exchange softening backwash water may cause concerns similar to membrane concentrate, depending on the concentrate water quality. Local regulatory agencies should be consulted for methods to dispose of ion-exchange backwash.

WASTEWATER TREATMENT WASTE DISPOSAL

All of the process wastes described in the previous section of this book may occur in wastewater systems, but the major constituent for disposal is biological sludge generated with the aeration process. Sludge is the dead and dying bacteria from the treatment process—not solids from the influent. Sludge disposal is becoming increasingly problematic as a result of pathogen and nutrient runoff concerns. Most sewer plant sludges are easier to dewater than their water plant counterparts because the water plant sludge-producing processes tend to hold water. The exception is when alum is used for phosphorus removal; alum sludge is particularly hard to dewater.

Most large facilities dewater their own sludge and dispose of it in landfills, through land spreading, or via treatment that permits its use as fertilizer (Milorganite® is the most famous example). Land application is the most common disposal method, but the sludge rules require that pathogens be eliminated if public access is permitted. Regulations significantly limit the type of crops or livestock that are permitted on land application sites where pathogens are not eliminated.

This page intentionally blank.

Chapter 8

WATER DISTRIBUTION AND SEWER COLLECTION SYSTEM OPERATION AND MAINTENANCE

WATER DISTRIBUTION SYSTEM FACILITIES

A water distribution system is the collection of pipes, valves, fire hydrants, storage tanks, and reservoirs that carry water from the water source(s) or treatment plant and deliver it to customers. The system consists of large numbers of pipes, service lines, meters, some pumps, and limited storage reservoirs. Table 8-1 is an inventory of one utility's water distribution system. Where possible, this inventory should be updated continually and should include the age and material of pipe. Unfortunately, few systems have information this detailed, which makes long-term capital replacement planning more difficult.

WATER MAINS

The piping in the distribution system should be large enough to meet the maximum domestic and industrial use by customers, provide ample flow for fire protection, and allow for future expansion. Since fire flow is almost always the largest demand, it usually determines the pipe sizes required in the system. Occasionally, fire flow capacity cannot be provided due to economic reasons. This situation exists in some rural areas where homes are far apart and can only be served practically by small-diameter pipes that furnish only domestic needs.

Water mains are the lines in front of houses. Six-inch- and 8-inch-diameter mains are installed in most residential areas because they are considered the minimum size that will provide adequate fire flow. Looping of lines to eliminate dead ends is also important for fire flow purposes.

Transmission mains are the larger pipelines that serve a regional area. Transmission mains are installed to move large quantities of water from one point to another and are often connected to storage tanks, reservoirs, the treatment plant, or pumping stations.

Pipe Materials

The pipe materials used in the water distribution system vary from galvanized iron to asbestos cement to polyvinyl chloride (PVC) or ductile iron

Table 8-1 Inventory of water distribution system

Item	Number of Units	Units
2-in. and under water main	45,110	lin ft
4-in. water main	18,990	lin ft
6-in. water main	155,490	lin ft
8-in. water main	192,510	lin ft
10-in. water main	51,330	lin ft
12-in. water main	53,400	lin ft
16-in. water main	23,970	lin ft
18-in. water main	2,850	lin ft
20-in. water main	390	lin ft
24-in. water main	2,280	lin ft
12-in. raw water line	10,000	lin ft
⅝ or ¾-in. services	7,135	ea
1-in. services	175	ea
1-½-in. services	125	ea
3-in. services	2	ea
4-in. services	1	ea
6-in. services	3	ea
8-in. services	4	ea
Fire hydrants	744	ea

NOTE: lin ft—linear feet; ea—each

(see Figure 8-1), depending on the age of the system and location. Water distribution pipelines are perhaps the most neglected area of a water system. In many areas, the age of the water distribution pipes averages more than 50 years old. Local soil conditions and other factors should be considered to evaluate whether pipes more than 50 or 60 years old exceed their useful life. Pipe age is especially critical if, for portions of the year, the pipes are partially submerged. An investigation of the condition of pipelines in submerged conditions should be undertaken periodically (including comments recorded during repairs) to evaluate the state of deterioration of older pipelines and the priority for replacement. The situation with pipes submerged in salt water is especially acute.

Failures of these pipelines, especially large ones, potentially will cause road and property damage as well as service disruptions, so a proactive approach is needed.

Cast-iron pipe has been used for water mains for many years; some cast-iron pipe is more than 100 years old and still provides good service. The use of cast-iron pipe diminished in the 1950s when it was replaced with ductile-iron pipe. Many water systems built with Works Progress Administration funds during the Great Depression were made of cast iron. An ongoing issue with these older cast-iron pipes is that the joints were usually made with lead driven or poured

Figure 8-1 Ductile iron and C900 PVC pipe

into the joint to seal it. Periodically, the lead pops out and the pipes leak. To repair these lines, lead joints must be driven back into the joints and/or bell clamps installed. Lead joints in high-traffic areas are particularly susceptible due to truck vibrations. Cast iron is brittle under certain loads or freezing.

Older cast-iron pipe had no lining, and under some water conditions, moderate to severe tuberculation occurs in the pipe. Tuberculation involves the buildup of rust/bacteria in mounds on the pipe interior. Tuberculation reduces the capacity of the pipe both because of constriction of the opening and the added roughness of the walls. When tuberculation seriously restricts flow, the main must be mechanically cleaned or replaced.

Ductile iron is the newer version of cast iron that is now in general use. The material appears to be about the same as cast iron, but it has been treated so that the metal is somewhat flexible and therefore less subject to breaking or cracking. Most ductile-iron pipe is delivered with a cement coating to prevent interaction of water with the iron. Ductile joints can be mechanical but are usually push-on joints with rubber gaskets. The lining eliminates the tubercle problem, and rubber gaskets replace the lead ones. American Water Works Association (AWWA) Standard C104 address ductile-iron pipe coatings.

Asbestos–cement pipe is made of cement mixed with asbestos fibers for reinforcement (AWWA Standard C400). It was initially used extensively in the 1960s because of its light weight, ease of handling, competitive price, and relative resistance to corrosion. However, when disturbed, asbestos–cement pipe can become brittle and shear under certain soil or installation conditions. Also, concern about asbestos fibers when sawing the pipe has dictated special procedures for workers. Many water systems no longer permit asbestos–cement pipe, and others are planning long-term replacement programs to remove it from service. AWWA has also withdrawn all standards, including C400.

Polyvinyl chloride water main pipe has been used extensively in many water systems since the 1980s. PVC is light, flexible, easy to cut and work with, color coded, competitive costwise with ductile iron, and does not appear to deteriorate in the ground. Pipe sizes make it possible to connect with ductile iron directly. AWWA has created standard C900 to address PVC water distribution piping to ensure pipe quality. Thinner-walled pipe often does not provide the same life as thicker C900 PVC. C900 PVC is available in many sizes. C905 is the AWWA standard for PVC pipe larger than 12 inches.

Many water systems specify blue pipe for water (green for sewer and purple for reclaimed wastewater). PVC should not be stored outside for any period of time; sunlight quickly degrades the pipe's strength. Discolored pipe should be rejected on construction sites for new installations and discarded from utility parts yards. Well-bedded PVC provides a life similar to ductile iron although it can be crushed more easily than ductile- or cast-iron pipe.

Galvanized iron was a common material for small-diameter pipelines (3 inch and smaller) and service lines in the 1950s and 1960s. Galvanized iron is simply iron pipe with a metallic coating of zinc. Damage to the zinc coating provides a point for corrosion to start. Experience throughout the Southeast indicates that the acidic soil conditions do not promote long life of galvanized pipelines, so these small pipelines may be significant sources of leakage. They also may contribute to water quality problems as a result of deterioration of the iron from aggressive waters. Replacement of these pipelines should be a priority as sandy or clayey soils are not conducive to indicating small leaks. As a result, these galvanized lines could be a source of significant leakage on the system.

Larger water mains are also constructed of reinforced concrete, steel, or fiberglass depending on their intended use, type of installation, and soil conditions. AWWA publishes installation and materials standards that should be followed for each type of pipe and fitting to ensure long, low-maintenance life for field crews. Prestressed concrete pipe (C303) and steel are used for large transmission piping but not for local water mains.

Depth of Pipe

The critical concerns in burying pipe are the prevention of freezing, the minimizing of temperature fluctuations (hot or cold), and protection of the pipe. As a result, the depth at which water mains are buried varies greatly throughout the United States. Water mains can be buried quite shallowly in warm southern states (minimum 30 inches in most cases) because the only concern is physical damage. Mains are buried deeper where there is moderate ground frost and may be buried up to 8 feet deep in northern states where temperature changes and frost may expand and contract the pipe.

Figure 8-2 Broken fiberglass pipe

Pipe Problems

Pipe problems involve five areas: shear breaks due to freezing and differential settling (all piping), pipes that burst (typically from water hammer, as in Figure 8-2), tubercles that restrict water flow, corrosion (iron and galvanized), and excessive age. Water crews are expected to be able to fix small pipe breaks. To accomplish this, an appropriate inventory of materials must be kept on hand. Large pipe breaks may require contractors with specialized equipment.

Valves (AWWA Standard C500) and Hydrants (C502 and C503)

Valves are installed at intervals in water main piping so that segments of the distribution system can be shut off for maintenance or repair. Valves should be located close enough so that only a few homes or businesses will be without water while a leak or break is being repaired. Valves should be installed in a normal grid system (see an example in Figure 8-3) in which mains are in all the streets and run in every direction of the grid. It is recommended that at least three valves be installed at each intersection.

Each valve should be installed with a valve box that extends to the ground surface and has a cap that can be removed so that a valve key can be used to operate the valve. Valves should, if possible, be located where the box is easily located and where damage by snowplows and other equipment is least likely.

A valve exercise program should exist to ensure that valves are working by opening and closing them at least once per year. Water distribution system valves should be periodically operated because this provides an opportunity to ensure that valve boxes are exposed and have not been filled accidentally with

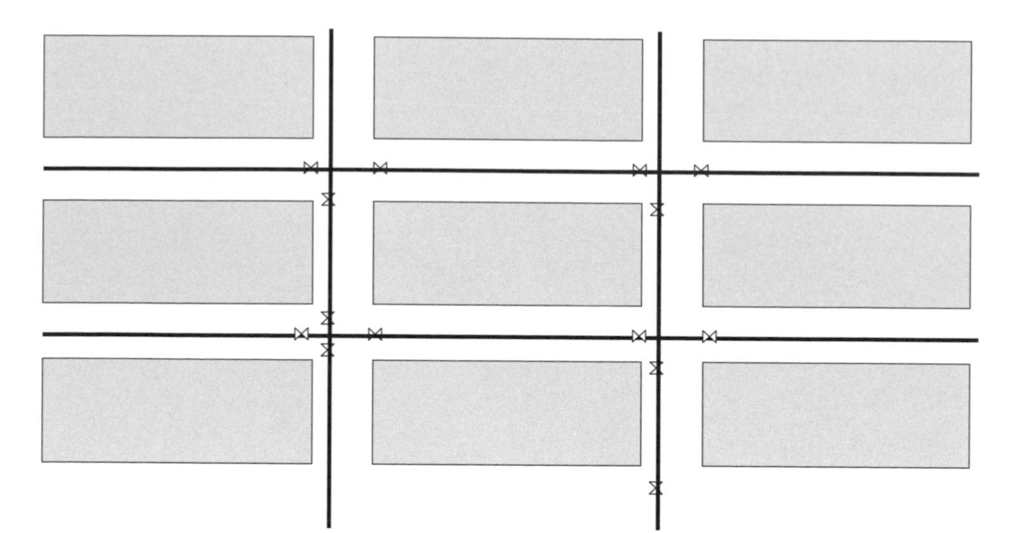

Figure 8-3 Water mains and valves installed in a grid pattern

dirt or damaged by paving or snow removal activities. A valve exercise program ensures that the valves are open and work properly, and it loosens up the valves so that they will operate more easily. Systems with a large number of valves often purchase power valve-turning equipment to speed the job of exercising their valves. When valves are operated, the number of turns should be counted to make sure they are fully operated in both directions. Valves that do not operate properly should be dug up and repaired as soon as possible.

Leaking or damaged valves should be replaced when discovered to minimize outages. Valve exercising programs may be the most ignored program for water systems; very few systems actually exercise valves, because few local officials see the importance—until a major break occurs, none of the necessary valves work properly, and large sections of the system must be shut down to make repairs. A typical valve is shown in Figure 8-4.

Fire hydrants are of two general types. A wet-barrel hydrant is full of water at all times and can only be used in parts of the country where there is no danger of freezing. Dry-barrel hydrants have the valve located at the bottom of the barrel and are operated by a long shaft extending down from the operating nut on the cap. Dry-barrel hydrants also have a small valve connected to a weep hole at the bottom that allows water to drain from the barrel when the hydrant valve is shut off (see Figure 8-5).

Hydrant locations should be selected carefully. Hydrants should be readily visible and located near a paved surface where they will be accessible by firefighting equipment. They should also be placed where they are protected from damage by vehicles and are least liable to be covered by plowed snow.

Public officials should always insist that police enforce parking restrictions adjacent to fire hydrants so that hydrants will not be blocked if needed. Police should also be reminded to watch for vandalism and unauthorized use of hydrants and to report incidents to the water system manager. Frequent painting

Figure 8-4 Valve

with bright paint protects hydrants from rusting and makes them easy for the fire department to find. Well-maintained hydrants also project a positive public image of the water system. In some systems, fire hydrants are color coded by flow volume.

Water Main Upgrade Programs

Water main upgrade programs are designed to replace current pipelines—those that are small, galvanized, or provide insufficient service and/or no fire protection. Typically, the water utility replaces these small lines with 6-inch or 8-inch pipelines made of PVC C900 or ductile iron, to provide fire protection. Leaky pipes and salt-immersed pipes should be a priority for replacement with appropriate pipe materials.

PVC C900 and ductile-iron pipelines generally cost under $50 per foot for small pipelines (6 to 8 inches in diameter). Many can be installed by water system crews. Other pipeline maintenance programs should address low-flow or low-pressure problems in the system. Existing developed areas are the first priority. Another priority should be completing loops that will address water distribution system pressure issues. Both of these programs have the benefit of increasing water sales to the areas where the loops are made or the lines upsized since flows are no longer restricted.

WATER SERVICE PIPES

The small-diameter pipe used to carry water from the water main connection to an individual building is referred to as a water service pipe (Figure 8-6). A water service pipe may range from 3/4 inches (20 mm) in diameter for a small home to 6 inches (150 mm) for an apartment building. Large buildings and industries often have service and fire sprinkler pipes that are even larger. AWWA

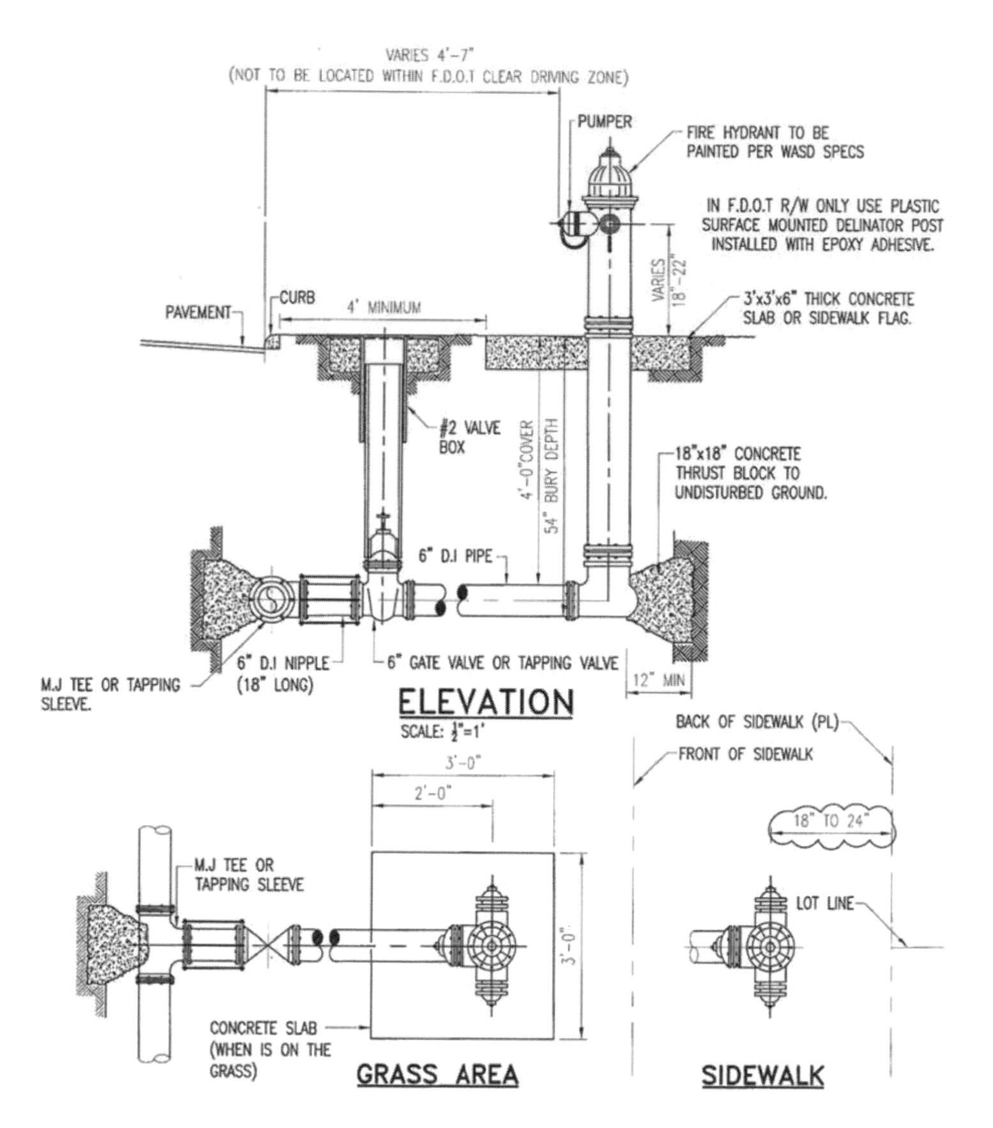

Published with permission from Miami–Dade Water and Sewer Department.

Figure 8-5 Typical fire hydrant installation

Standard C800 was designed for service lines (C901 and C903 apply to plastic pipe).

Each water service pipe usually has a buried valve called a curb stop inserted in the line at a point at the edge of the public street or alley right-of-way or an easement. Where curbs and sidewalks exist, water system policies generally standardize the curb stop location at a set distance between the curb and sidewalk or at the lot line. The buried valve is fitted with an adjustable service box (curb box) that extends to the surface and has a removable cap so that a valve key may be inserted to operate the valve. The curb stop is primarily used to shut off the

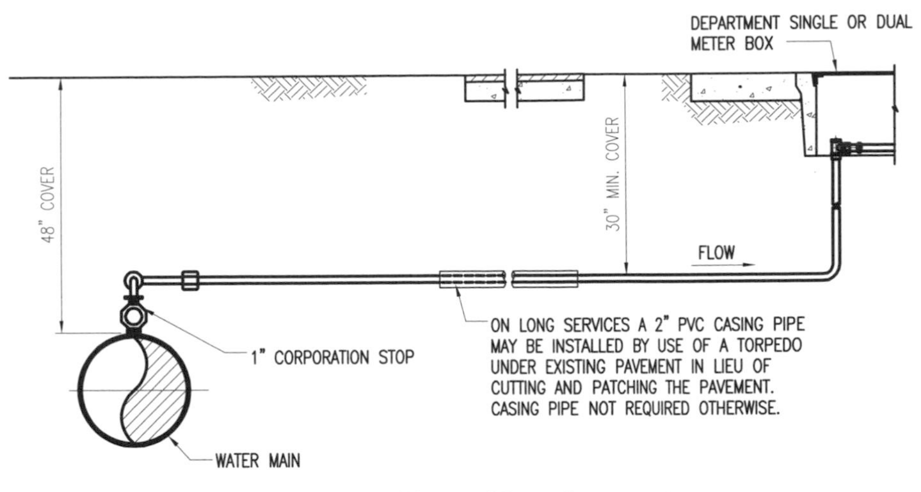

Published with permission from Miami–Dade Water and Sewer Department.

Figure 8-6 Typical water service installation

service if the building being served is vacant or repairs are needed. It is also a way of discontinuing service for nonpayment of the water bill. Local ordinances usually address turn-off of services for nonpayment.

AWWA has the following policy statement (2008a) of service discontinuation for nonpayment, most recently approved by the Board of Directors on June 8, 2008:

> AWWA realizes the importance of the nondiscriminatory billing and collection procedures to ensure that each customer pays for the services rendered by the utility under its rates and tariffs. Failure on the part of the customer to pay a water bill necessitates that other customers bear the burden of paying for the service.
>
> AWWA recognizes that certain circumstances may require some flexibility because water service is a necessity in maintaining sanitary conditions in the home, and may be required for life-sustaining equipment. It may also be a vital part of industrial and commercial operations. Discontinuance of water service for nonpayment is considered a final phase of a collection procedure and should be instituted with sufficient notification when all other reasonable alternatives have been exhausted.

Maintenance Responsibility of Water Services

Water system policy varies on responsibility for maintenance and replacement of water services. Some systems require that all of the service, beginning at the main, be maintained by the property owner. Other systems require property owners to maintain only the portion beyond the curb stop, lot line, or meter pit. The utility system policy should be made clear by ordinance or approved policy resolution.

Pipe Materials

Water service pipes are generally made of lead, galvanized iron, copper, or plastic. Lead was the best material available for small pipes when the first water systems were developed. Lead water service pipes are still in use in many older systems. Lead service lines are relatively flexible and resist corrosion but gradually become more likely to leak or break as they get older. Lead is no longer used for service lines, but many lead pipes may still exist in water systems.

New federal regulations, designed to protect the public from the danger of lead in drinking water, require systems to ensure that leaching of lead from water services is minimized. Systems with aggressive water that tends to dissolve lead may have to install additional chemical treatment to meet the requirements. Systems that cannot adequately control the leaching of lead may be required to remove existing lead service pipes and replace them with other material.

Galvanized-iron pipe, used for water service piping for many years, corrodes very quickly in some types of soil. This material should be replaced since it is a source of leakage in the system. The same issues apply for galvanized service lines, except that the service lines are smaller and tend to deteriorate more quickly than galvanized mains.

Copper pipe came into use in the early 1900s and gradually became the preferred material in many parts of the country. Copper is flexible, fairly easy to install, resistant to corrosion, and lasts almost indefinitely under most water and soil conditions. However, copper can also leach from the service to water supplies where the potable water is aggressive. Lead solder was also used until the mid-1980s for connecting copper pipe. This lead solder was noted as a source of leaching lead in water systems. As a result, it is no longer used in the United States for copper service line solder.

Plastic pipe has been used for water services since shortly after World War II. Polyethylene and polybutylene are the common materials. Both are lightweight, easy to install, flexible, moderately priced, and resistant to corrosion. In some areas, plastic pipe has been used almost exclusively for years. There are many types of plastic, but only certain types and grades are approved for potable water use. AWWA standards define these materials:

- ANSI/AWWA C901 Polyethylene (PE) Pressure Pipe and Tubing, ½ In. (13 mm) Through 3 In. (76 mm), for Water Service
- ANSI/AWWA C903 Polyethylene–Aluminum–Polyethylene and Cross-linked Polyethylene–Aluminum–Cross-linked Polyethylene Composite Pressure Pipes, ½ In. (12 mm) through 2 In. (50 mm), for Water Service

Plastic pipe must be tested for durability and freedom from constituents that might cause tastes or odors or might release toxic chemicals. Only pipe that has the seal of an accredited testing agency (typically NSF International) printed on the exterior should be used for potable water purposes.

WATER STORAGE FACILITIES

The primary reason for providing storage of treated water is to have a reserve supply readily available for firefighting; emergencies such as repairs to treatment facilities or pumps, loss of water supplies, pipe breaks, or contamination; or for periods of heavy water use. For instance, the stored water can be used to maintain pressure for a period of time if a well or pressure pump should fail or lose power. Storage will also provide added capacity during a fire and help to temporarily maintain pressure in the distribution system during a water main break. Another function of storage is to allow a treatment plant to operate at a relatively constant rate. When customers are using water at a low rate, excess water can be stored. When use is high, stored water can meet the demand and operation of the treatment plant will not have to be altered.

The quantity of water storage that should be provided on a system is usually based on the amount of water required to meet domestic and fire flow needs. Many systems find it advisable to furnish more than the minimum storage capacity. For instance, storage of enough water to last one or two days may be provided by systems that depend on a single long transmission main for source water. Systems that have periodic episodes of temporarily poor quality in their source water also frequently provide enough storage to allow them to avoid taking water until the source water quality improves.

Types of Storage Facilities

Water storage facilities fall into the categories of elevated tanks, standpipes, hydropneumatic systems, and ground reservoirs. Elevated tanks are the most familiar because they are visible in prominent locations in most communities. Elevated tanks are generally constructed of steel, with the tank portion supported on legs or a pedestal (see Figure 8-7). Tanks are generally located on the highest ground that is available and acceptable to the residents. The public is not generally bothered by an existing tank located in a residential neighborhood but usually will not want a new tank erected near their homes.

An elevated tank normally fills and empties in response to demands on the water system (referred to by operators as "riding the demand"), and the elevation of the water in the tank determines the water pressure on the system (see Figure 8-8). A signal that indicates the water level in the elevated tank is commonly used to vary the operation of the pressure pumps supplying the system.

When the water level is near the top of the tank, the supply of water is reduced or stopped before the tank overflows. When the water level falls to a predetermined point in the tank, flow to the tank is increased (Figure 8-8). Occasionally, a water system must operate at a pressure that would cause an overflow of the elevated tank. In this case, water is admitted to the tank by an automatic valve that shuts off flow before the tank overflows.

In water systems, a standpipe generally refers to an aboveground tank that is the same size from the ground to the top. Standpipes are primarily used where they can be located on a high point of land so that all or most of the stored water will furnish usable pressure to the water system (see Figure 8-9).

Figure 8-7 Elevated tank

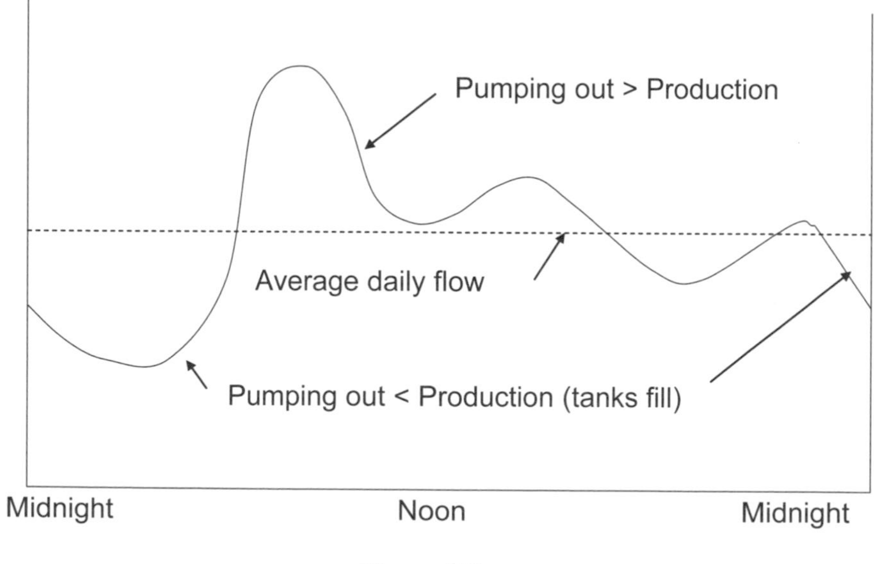

Figure 8-8 Typical filling and emptying cycles of a tank and reservoir

Hydropneumatic systems have been developed primarily to serve small systems where an elevated tank is not practical. A large pressure tank is buried or located aboveground and kept filled partly with water and partly with compressed air (see Figure 8-10). The balance of compressed air against the water maintains the desired pressure in the system and forces water out of the tank when needed. An air compressor is required to maintain the proper air-to-water ratio.

A water reservoir (see Figure 8-11), ground storage tank (see Figure 8-12), or wet well is generally a large tank in which treated water is stored under no pressure. The water must be pumped out of the reservoir and into the system when needed. Reservoirs are constructed of concrete or steel and may be aboveground, partially underground, or completely buried. Water is usually admitted to a reservoir by a remotely operated valve during times when excess water is available, such as in the middle of the night. Pumps are then operated to add water from the reservoir to the system as needed during the day or in an emergency.

Occasionally, a water system has a high point of ground available where a reservoir can be constructed so that it will supply adequate pressure to the system without the need for repumping.

The prime advantages of a reservoir are that it can be constructed to store relatively large quantities of water and can be completely buried where an aboveground structure would be objectionable to residents. When a reservoir is completely buried, the land above it is sometimes used for a park or recreational area. The prime disadvantage is the cost of power to operate the pumping equipment and contamination due to recreational use if uncovered.

Figure 8-9 Standpipe

Figure 8-10 Hydropneumatic tank

Figure 8-11 Water storage reservoir in West Palm Beach, Fla.

Figure 8-12 Ground storage tank

WATER METERING

Types of Meters

The principal uses of water meters on a water system are to record the amount of water treated and delivered to the water system and to measure water used by customers. The two general types of meters used are velocity meters and displacement meters.

Velocity meters are used when large quantities of water must be measured. They work on the principle of converting a measurement of the velocity of water passing the measuring point into quantity of flow. These meters are commonly used at each well or water intake, at intermediate points in the treatment system, and at the points where water enters the distribution system. Velocity meters are also used to measure the amount of water admitted to and/or pumped from reservoirs and may also be used for customers that use large quantities of water. Velocity meters commonly measure water flow by means of propellers (see Figure 8-13), turbines, pressure measurement, and electronic sensing. The meter register then automatically translates the registration into gallons (liters) or cubic feet (cubic meters).

Displacement-type meters measure the number of times a container of known volume is filled and emptied. This method is more accurate than a velocity meter but is only practical for measuring relatively low flow rates. The nutating disk meter (see Figure 8-14) is a type of displacement meter commonly used for water services because it is durable, relatively trouble free, moderately priced, and quite accurate over the normal flow range of most water customers. Figures 8-15 and 8-16 are examples of water meter assemblies in service.

Figure 8-13 Propeller meter mounted directly in pipeline

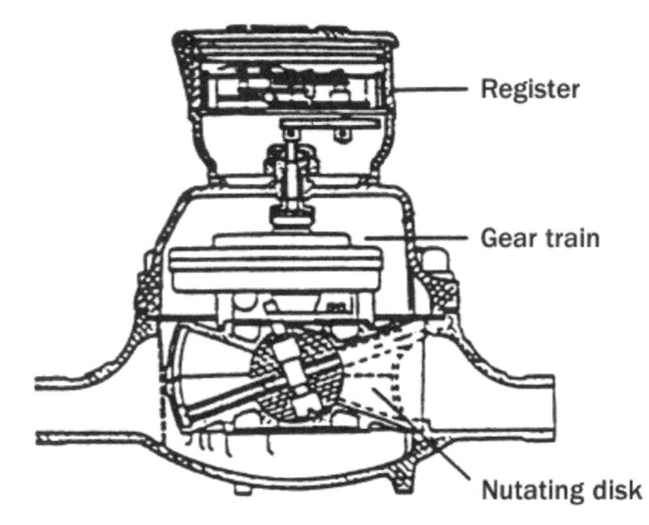

Figure 8-14 Propeller meter mounted directly in pipeline

The following AWWA standards are available for meters:

- ANSI/AWWA C700 Cold-Water Meters—Displacement Type, Bronze Main Case
- ANSI/AWWA C702 Cold-Water Meters—Compound Type
- ANSI/AWWA C704 Propeller-Type Meters for Waterworks Applications
- ANSI/AWWA C710 Cold-Water Meters—Displacement Type, Plastic Main Case

Figure 8-15 Actual meter

Figure 8-16 Meter installation: fire meter and bypass meter

Water Production Metering

The rate of water flow from pumps, wells, and treatment facilities in the United States is usually expressed in gallons per minute (gpm). Flow is expressed in liters per minute (L/min) and liters per second (L/sec) in the rest of the world. A well pump, for instance, may be rated to produce 500 gpm (1,900 L/min or 31.7 L/sec). The total output of a water plant or system is usually expressed as million gallons per day (mgd) in the United States and million liters per day (ML/d) elsewhere. The total quantity of water pumped to the distribution system should be accurately measured and recorded so that it can be compared with the water sold. The difference between the water pumped and water sold is the amount of water unaccounted for—that is, the amount that is lost to leaks or otherwise lost and unpaid for in the distribution system. Water systems should strive for an unaccounted-for fraction of less than 15 percent. Analysis of production meter records should also be made to check such things as well productivity, trends in customer use, and the need for system expansion.

Customer Metering

All water systems should meter their customers, and everyone who uses water (or sewer) should be deemed to be a customer. The following position was adopted by the board of directors for AWWA on January 26, 1969; revised June 15, 1980; reaffirmed June 22, 1986; and revised June 6, 1993, June 21, 1998, and June 18, 2004 (AWWA 2004a):

> The American Water Works Association (AWWA) recommends that every water utility meter all water taken into its system and all water distributed from its system at its customer's point of service. AWWA also recommends that utilities conduct regular water audits to ensure accountability. Customers reselling utility water—such as apartment complexes, wholesalers, agencies, associations, or businesses—should be guided by principles that encourage accurate metering, consumer protection, and financial equity.
>
> Metering and water auditing provide an effective means of managing water system operations and essential data for system performance studies, facility planning, and the evaluation of conservation measures. Water audits evaluate the effectiveness of metering and meter reading systems, as well as billing, accounting, and loss control programs. Metering consumption of all water services provides a basis for assessing users equitably and encourages the efficient use of water.
>
> An effective metering program relies upon periodic performance testing, repair, and maintenance of all meters. Accurate metering and water auditing ensure an equitable recovery of revenue based on level of service and wise use of available water resources.

Meters should be sized properly according to their use. Oversized meters will underregister usage; undersized meters will also underregister due to wear while large meters will tend to lose calibration within 24 months of installation. Therefore, it is important that all water systems establish a water meter repair or replacement program to ensure that meter readings are as accurate as possible and that unaccounted-for water is minimized. Large meters should be removed for testing and repair in a meter shop maintained by the water system or sent to the factory or an outside firm to be reconditioned no less often than every two years. The economics of disposing of worn meters and replacing them with new meters should also be considered. Small meters are rarely worth fixing. Table 8-2 shows the appropriate sizing for water meters depending on the units, plumbing fixtures, or usage anticipated.

Meter Reading and Billing

Meter reading is usually housed in either the water distribution or finance section of the utility. Customer meters are read at regular intervals by a meter reader who travels from house to house. Readings for each address are entered on a meter card, in a meter book, or in a handheld meter-reading recording device. Readings are then submitted to the billing office, usually housed in a finance department, to begin the billing process. Many small systems place multiple meter readers in the field at the same time to get all meters read within a few days, which allows for all billing on a set date. In larger systems, meter reading is usually on a continual basis.

In warm climates, domestic water meters may be located in a garage, buried in a shallow pit, or exposed near a building. In areas where there is danger of freezing, meters are most commonly placed in customers' basements. Where there is no basement, some systems will allow the meter to be installed in a ground-floor utility room or closet. Meters are also frequently installed in pits in the right-of-way where there is not a suitable location in the building. Some municipal systems and most rural water systems install meters in pits to avoid having to enter customer buildings and to facilitate meter reading and repair.

Table 8-2 Meter size chart

Water Meter Size (in.)	Meter Type	Maximum Fixture Units	Peak Maximum Volume (gpm)
⅝ × ¾	Displacement	25	20
1	Displacement	45	50
1½	Displacement	100	100
2	Compound	225	160
3	Compound	500	320
4	Compound	750	500
6	Compound		1,000
8	Compound		1,600
10	Turbine		2,900
12	Turbine		4,300

In northern areas, some or most meters may be located inside buildings. As a result, there frequently may be a problem in getting the meter reading from homes where the residents are not home on weekdays or are on vacation. Some systems partially solve this problem by having meter readers work on Saturdays. Other systems ask the meter reader to leave a message on the door if nobody is at home. The card requests the occupant to read his or her own meter and return the attached postcard by a specified date.

When a meter reading cannot be obtained, the bill is estimated, based on use in preceding periods and the time of year. Estimated readings can only lead to dissatisfaction from the customer, so it is generally not good policy to estimate a bill more than once or twice before an actual reading is obtained.

Meter readers usually receive a better reception for entering residences if they are provided with neat distinctive uniforms and proper identification. It also helps if readers always work the same area so that customers recognize them.

Several methods of remote meter reading devices have been developed to speed meter reading and reduce customer inconvenience by allowing an inside meter to be read from outside the building. Other remote reading devices have a special transmitter mounted in the meter that sends a signal through a wire to a register mounted on the exterior of the building. Some units have a reproduction of the meter register on the exterior. Others are designed to allow the meter reader to plug in a portable unit that electronically records the meter reading for later interpretation by a computer.

SEWER COLLECTION SYSTEM

The materials used in sewer collection systems fall into two categories. Most sewer collection systems are gravity systems, whereby the wastewater flows downhill to a treatment facility or collection point. For many utilities it is not possible to have wastewater flow all the way to the treatment facility, so at the low points, a lift or pump station is installed. The lift station is a large manhole with pumps in it (see Figures 8-17 to 8-19). The waste leaving a lift station is under pressure. For these systems, the piping materials are the same as for water distribution pipe: cast or ductile iron (lines with cement or polyethylene), PVC, or asbestos–concrete. Valves and fittings are also the same as for water distribution systems.

For the gravity collection system, the piping system has traditionally been very different. The collection system consists of the gravity pipes, manholes, service lines, and cleanouts. The manholes and cleanouts are required for access and removal of material that may build up in the piping system.

Collection System Piping

Collection system piping throughout North America prior to 1980 was predominantly vitrified clay. Since that time, asbestos–concrete and various grades of PVC have been used. Ductile iron is rarely used due to the potential for corrosion from hydrogen sulfide gas.

Vitrified clay pipe has been used for well over 100 years. The pipe is resistant to deterioration from virtually all chemicals that could be in the water and from soil conditions. It has a long service life when installed correctly and left

Figure 8-17 Lift station in sewer collection system

Figure 8-18 Lift station control box

undisturbed. However, vitrified clay pipe is brittle, so settling from incorrect pipe bedding, surface vibrations, or freezing can cause the pipe to crack. There are also limitations on pipe size. Temperature differences between the warm wastewater and cooler soils can cause the exterior pipe surface to be damp. The dampness encourages tree roots to migrate to the pipe, where they may wrap around the pipe. Where cracks occur, roots will enter the pipe. Over the long

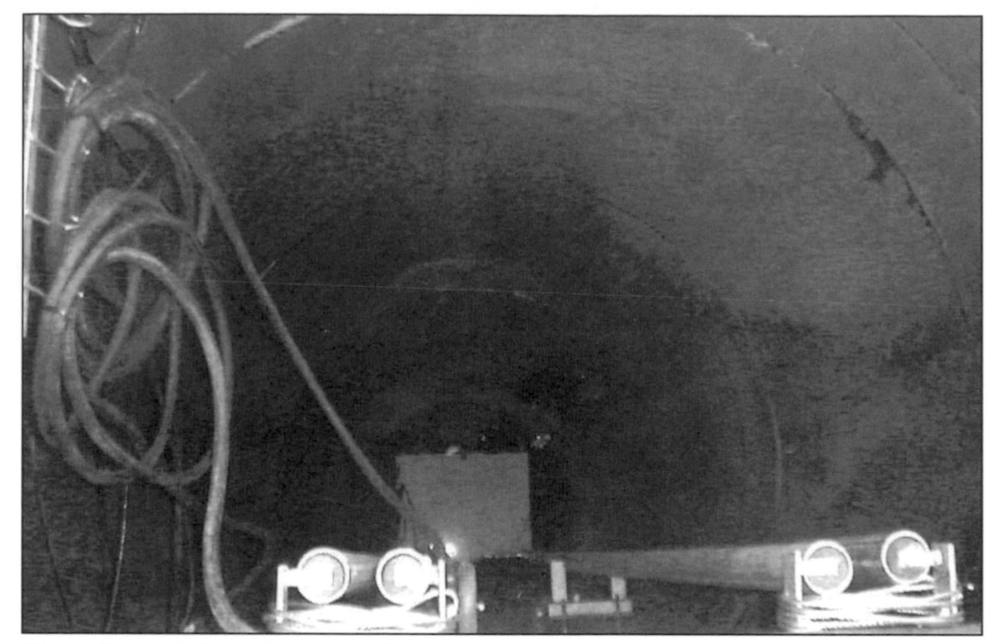

Figure 8-19 Lift station wet well

term, the pipe will become broken and damaged from the roots, vibrations, and freezing. Where the water table is above the pipe level, significant infiltration can occur, which reduces the capacity of the wastewater treatment plant.

A second concern with older vitrified clay pipe is the short joints used—as small as 2 feet prior to 1920, 4 feet prior to 1960. Field joints were made prior to 1920 and even later. The joints were sealed with cement and cloth "diapers" wrapped around the joint. However, concrete is not waterproof and will crack with time. The combination results in piping with many joints, each of which has the potential to leak. Even today, the joints are short compared to PVC and ductile iron (20 feet and 18 feet, respectively), although the joints and material have improved substantially. Vitrified clay remains the choice of material to use in industrial areas where pipe protection is required. Lining vitrified clay pipe is possible with many products, thereby extending the life of the pipe.

Concrete pipe is used for large-diameter sewer lines. Concrete has the benefit of durability and structural strength. Concrete piping as large as 96 inches is not uncommon. Concrete is the only material made in these diameters (usually the pipe is pre-stressed when this large). Concrete piping suffers one significant problem—its vulnerability to hydrogen sulfide. As sewage remains in the piping system, if air is not entrained, the sewage will become septic. Septic sewage is black and smells heavily of hydrogen sulfide (rotten eggs). Hydrogen sulfide, when attached to pipe surfaces, reacts with water to form sulfuric acid, which is deleterious to the pipe. The area just above the water line and the crown of the pipe are most vulnerable. Without proper inspection, maintenance, and care, the pipe will fail. Lining the pipe is beneficial, but the lining must be monitored to ensure it remains. High levels of hydrogen sulfide can reduce pipe life by 20 or more years if left unchecked.

There are two ways to reduce hydrogen sulfide levels—chemical additives and aeration. If the sewage is not allowed to become septic, a difficult thing to do in large systems or where significant pumping is involved, hydrogen sulfide is not produced. Chemical additives can be put into lift stations and manholes to react with the hydrogen sulfide as well. These can be expensive.

Ductile-iron piping used in sewer systems is similar to that for water distribution piping. There is one area of difference: wastewater piping is usually lined with polyethylene instead of cement to prevent hydrogen sulfide deterioration. Ductile iron, like concrete, is susceptible to hydrogen sulfide corrosion.

A variety of PVC grades have been used for sewer lines, including C900 and SDR 35. More utilities appear to be favoring the heavier C900 pipe as a result of concerns during construction (bedding particularly, and backfill methods). PVC pipe is excellent for residential services, but the longevity of the pipe rests on having proper bedding and backfill methods. Improper backfill will crush, crack, or deform the pipe. C900 is more resistant to damage, and the increase in cost is minimal. All PVC sewer lines should be green.

Manholes

Manholes are access sites for workers and for changes in direction of the pipe. Manholes are traditionally precast concrete or brick. Brick was the method of choice until the 1960s. Brick manholes suffer from the same problems as vitrified clay sewer lines—the grout is not waterproof, so the grout can leak significant amounts of groundwater. The manhole cover may not seal perfectly, becoming another source of infiltration (see Figure 8-20). Precast-concrete manholeslimit the number of joints. Elastomeric seals are placed between successive manhole rings. Many utilities require the exterior of the manholes to have coal tar or epoxy covering the exterior, which helps to keep water out.

Service Lines

Service lines are made of the same piping as the collection system. Services lines are generally 4 inches or 6 inches. Schedule 40 PVC and vitrified pipe are the most commonly found materials, although Orangeburg pipe may also occur on older service lines. Service lines may be significant sources of infiltration due to roots and breakage. Unfortunately, the majority of service line length is on private property. Requiring home owners to repair these pipes is often politically difficult.

OPERATION AND MAINTENANCE OF PIPING SYSTEMS

Pipe System Records

Operators who work for a small utility for a long time often keep a mental map of the system. While valuable to the utility in the near term for responding to emergencies, to protect the system in the long term, especially in light of potential emergency situations, all records should be put on paper and/or computer file for later reproduction, modification, and improvement. Elected officials should recognize this problem and arrange for staff engineers or consultants to record the system information onto permanent maps and records.

Every utility system must maintain an up-to-date water distribution and sewer collection system map. It is surprising how infrequently such maps are available. Many water systems maintain the official distribution and collection system map as a series of maps in a single plat book kept in the engineering or utility administration system office. An official copy is needed, but copies should be available for field crews who are more often the people looking for the information on the maps. Computer disk and microfiche records for field crews may have some benefits over paper copies, which may get wet.

The water distribution map should include pipe size; valve, hydrant, and storage locations; and information on pumping systems. Fire hydrants should also be shown on the detailed records with measurements for both the hydrant and the hydrant control valve.

The sewer collection system map should include the location and size of gravity collector mains, force mains, lift stations, manholes, and valves. More detailed records should track information such as measurements from each valve to aboveground features (such as trees, curbs, and extended lot lines) in the event the valves are buried during street paving activities or disturbed by snowplows. Lift station records should include make and model of pumps, operating conditions, and pumping rates. Pipeline information should be as detailed as possible with indications of size, type of material, depth, periodic location measurements, and date of installation where possible.

Service line information should be recorded as installations or repairs are made. A metallic water service pipe with an inadequate location record can

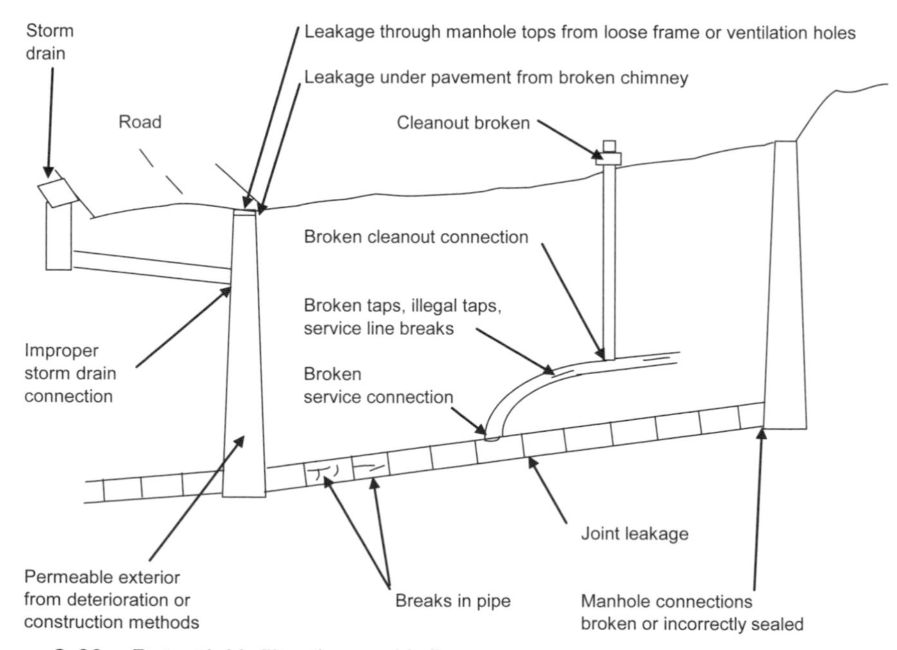

Figure 8-20 Potential infiltration and inflow areas

usually be located using an electronic pipe locator, but the process takes longer than using good record information. A pipe locator does not generally work on plastic pipe, so finding a pipe with no record can be difficult unless metallic tape was also installed (the normal practice).

Water distribution and sewer collection system maps and other information can be maintained on geographic information system (GIS) or other computer systems. Information that should be obtained and recorded before the pipe is covered in the trench are measurements locating the water main tap, the type and size of pipe, burial depth, measurements to the curb stop or meter pit, location of the pipe at various points, and the location where the pipe enters the building. Photographs are always useful, especially digital photos that can be put into report files and stored on CDs.

A computerized work-order tracking system should be utilized to record pipe system repairs and the condition of both the interior and exterior of pipe and appurtenances. Whenever a piece of pipe is removed, information should be recorded about the pipe, location, and conditions of the soil, and a coupon should be photographed. This information should be compiled for later use when assessing water main conditions, potential for failures, tubercles, and wear. Some utilities may wish to keep a small piece of pipe tagged with the date and location.

In a perfect world, the utility will have complete, accurate system maps; detailed maps to locate all piping (with pipe size, installation date, and material), valves, and fire hydrants; details on all repairs made and the conditions surrounding those repairs; pipe condition reports; data reports on all new pipes installed and conditions found during excavation; and a spreadsheet file of all fire hydrants and valves, with a number assigned to each and records for each that include installation date, type, repairs, results of hydrant flow tests, exercise or flush dates, and so on. Rarely does a utility have all of this data or the ability to obtain it. However, good record keeping started as soon as is practical will build a base of knowledge in only a few years that will improve the efficiency of repairs. Staff must be assigned to perform the record keeping, data entry, and mapping.

Maintenance Needs

Water distribution and sewer collection system equipment is often neglected, since most of the valves, pipes, and other equipment are buried and seldom thought about. Some of the problems that are caused by poor system maintenance include customer complaints of poor water quality or lack of pressure, difficulties in repairing water main leaks, and inadequate or unreliable water availability for fighting fires. Emergencies, such as the events of September 11, 2001, show the need for all portions of the distribution and collection systems to be in good working order to minimize response time to emergencies and to restore service as soon as possible. Storms highlight the need to reduce infiltration and inflow into the collection system so as not to overwhelm the piping system and cause plant damage or sewage overflows into streets.

Properly installed water mains and sewer collection piping should perform well over a long period of time. However, they will, with time and environmental conditions, break and leak. Many breaks are simple shear cracks at right angles to the length of pipe and can be repaired by exposing the pipe and slipping a repair sleeve over the break. Water leaks should be repaired as soon as is practical to minimize risks to property, roadways, and the public health (from contamination). However, leaky pipes can rarely be repaired while under pressure.

For pressurized pipes, it is desirable to minimize the shutdown area when a leak occurs. Valves are used to turn off the flow of water to the leak area. Three conditions are required for this to be successful: the appropriate number of valves, accurate valve location records, and working valves. If the valves in the vicinity do not work, larger areas will be affected. Where valves cannot be located or are found to be inoperable, the repair crew must continue searching until an appropriate combination of valves is found, which wastes time. If the utility system has good records and a valve exercise program, it should be able to locate and operate the correct valves quickly to shut down the smallest possible section of main and progress with the repair.

Pipe Incrustation and Sediment

The buildup of rusty sediment in the bottom of mains is a common problem in water distribution. This sediment in water mains can be caused by iron or manganese that was in the source water and has precipitated in the pipe, or from the presence of iron bacteria in the system. It can also be from rusting of old cast-iron pipe. If the problem is not too severe, some systems have found that a thorough flushing of the system once or twice a year is sufficient. Dead-end water mains need to be flushed more often. While the sediment is generally not harmful to health, it may cause a few customer complaints when disturbed. If water flow is suddenly increased, such as from use of a fire hydrant, the sediment will be disturbed and can turn customers' water slightly rust colored or even dark brown. Customers will be reluctant to drink and use the water, and the water may badly stain laundry items. Where the problem is ongoing and severe, professional advice should be obtained on the best method of correcting the condition. Residents should always be notified if flushing is planned.

Old water mains that become encrusted to the point of seriously restricting flow can be cleaned using a power rodding machine or by pushing a flexible pig through the line using water pressure. In most cases, the incrustation will quickly reform on the interior of a cleaned pipe, so consideration should be given to applying a cement lining on the pipe interior or repairing an existing, damaged cement lining.

Flushing

Flushing is more appropriate for water systems, although sewer lines need cleaning as well. All water systems should flush the piping at least once per year to remove sediment and stagnant water. When flushing the system, residents should be notified so that they are not alarmed if the water appears murky. The chlorine residual should be increased to help retard biofilm growth in the pipelines. If significant problems are found, pigging is an alternative. Flushing the system gives operators the opportunity to test fire hydrants (the blowoff points), work valves, and detect areas where valves may be shut or otherwise not working.

Fire Hydrants

Fire hydrants should be operated and tested at regular intervals because many small systems can also go for years without having to use a fire hydrant for fighting a fire. Many systems have a yearly program that is a combination of flushing water mains and testing hydrants. Records should be kept of the static pressure, flow test results, and any repair work performed on each hydrant. In freezing climates, whenever a hydrant is used, an inspection should be made to make sure that the barrel has fully drained. Hydrants provide the water system with the ability to check flow and pressure, flush the system, and find closed or broken valves as a part of the valve exercise program. The three most common reasons for a hydrant not functioning properly are unnoticed damage, improperly drained or frozen barrel, and closed valves in the distribution system. All of these problems can be readily identified and corrected by a regular program of hydrant inspection and testing. Failure of hydrants to operate properly during a fire could leave water system personnel and managers liable to a suit for damages, in addition to causing needless risk to life and property.

UTILITY SYSTEM EXTENSIONS

Most utility systems assign the responsibility for water and sewer extensions to developers or others desiring the extension. Where the extension replaces or upgrades an existing line, the utility may undertake the extension on its own or in concert with a contractor. If the extension is relatively short, the work can often be done by utility system employees. For larger extensions, a contractor is usually employed. Employing plans prepared by a professional engineer and obtaining permits are always required prior to construction.

It is important that the engineer's specifications and manufacturer's recommendations be followed for proper bedding, blocking, and installation of the pipe, valves, and hydrants. A poor installation job can create extra work for years, repairing reoccurring leaks and broken mains. Water system personnel should inspect all pipe installations to ensure that appropriate construction and installation practices are followed to minimize future maintenance.

This page intentionally blank.

Chapter 9

MANAGEMENT OF THE WATER SYSTEM

SERVICE DELIVERY MECHANISMS

Traditionally, utility systems fall into three classes: those operated by governmental entities (public-sector-operated systems), those operated by developers, and those operated by private companies (which tend to be holding companies that acquire smaller utilities and provide services under a corporate framework). Utilities operated by public entities are usually oriented toward providing high-quality, customer-service-oriented water and sewer service to their customers, with enough redundancy to minimize the risk of system failure. Developer systems are operated as vehicles to provide service to an area so that the land can be developed. Effective operation is not a primary goal of developer systems except to the extent that establishing an interim utility will prevent placement of a moratorium on development. Private, investor-owned systems are operated to provide service to customers in a satisfactory manner, while allowing the owners of the company to earn a profit. All three delivery mechanisms must constantly change their methods and programs to meet increasingly stringent regulatory mandates, directives, and regulations and to provide service to a user base that is increasingly conscious of the industry practices and requirements.

Public-Sector-Operated Systems

The majority of large utility systems that exist in North America are operated by governmental entities. Originally, many of these large systems developed from the central cities, a common occurrence in the Northern and Midwestern United States and Canada where the central cities may serve all of a metropolitan area. Public-sector-operated systems have developed over time, based on bonds sold in anticipation of customer growth and, in some cases, with contributions from the general funds of the government in an effort to encourage growth.

The benefit of a public-sector-operated system is that in many cases, the system is operated by a staff that reports to an elected body, and therefore users of the system have some say in the policies and practices of the system. In addition, governments can coordinate their utility policies and expansions in a manner that is most conducive to growth and can focus on long-term community needs in concert with transportation corridors. When development is ready to occur, utility and transportation networks are in place. The expertise and experience that exist within these systems often transcend jurisdictional boundaries. This same experience and expertise can be used to manage growth in corridors where these services are available and can be required as a prerequisite to development, thereby allowing expansion of the system by contributions-in-aid-of-construction at no cost to the users or taxpayer or the public entity.

There is also an economic development benefit from a public system. Rates and charges for new connections are regulated by an elected body, in compliance with the rather clear confines of case law and extensive litigation involving public bodies so as not to unfairly penalize either new or existing users of the system. Public-sector ownership provides the avenue to eliminate health and safety hazards resulting from improperly operated water and sewer systems through police powers derived under statutes and case law. These same statutes allow local governments to require connection to water and sewer systems when the systems are available, and to enforce these connections while protecting the current rate payers by guaranteeing that the expenses will be recouped via property liens.

One of the most important benefits derived from public-sector utilities is the ability of the local government to spread the costs of the utility in an equitable manner to all those who benefit from the existence of the utility. The existence of a water and sewer utility not only benefits online users, but also benefits developers, who own land they want to develop. Public systems are generally very strong in public relations aspects due to the visibility of the rates, charges, and ongoing construction and their impact on elected officials. This causes public-sector systems to be very responsive to requests from the public.

However, public systems often suffer in part from the regulations put in place by elected officials, staff, and the public under the intent of protecting the public, such as extensive personnel and purchasing regulations that make it difficult to respond to emergencies and general equipment breakdowns. As a result, public utility officials tend to build in excess redundancy to ensure that service does not fail; ultimately, failure to provide service affects elected officials, whose only response is the termination of utility employees. At the same time, close contact between the decision makers and the public does not always foster good local decision making, as good decision making may be ignored for political gains (rate increases are an example). Politically charged issues brought before elected bodies often create unintended consequences and new issues that frustrate the objectives of the utility system. System maintenance and issues of finances, rates, assessments, and impact fees are often difficult to have approved despite their relative importance in the continued growth of the utility system.

Regionalization

In some cases, the most economic, efficient, and responsible way to operate a utility is to integrate the operation and maintenance of a series of water and sewer systems into larger regional utilities and then expand the services of the utility effectively to use the same operating personnel, billing procedures, policies, and objectives for the extension of this service. These regional authorities are almost exclusively public entities in the United States. The blueprint can be found in state and local comprehensive plans and operation of large utilities. A regional system approach should create economies of scale for operations and maintenance of the system, produce a system that can secure lower interest rates and terms for financing when additional facilities are necessary, provide more homogenous treatment of rate payers, and ensure safer and more environmentally sound maintenance and operation of facilities and treatment of water. American Water Works Association (AWWA) and the US Environmental Protection Agency (USEPA) support regionalization efforts under the auspices of multijurisdictional public bodies.

Developer Systems

Developer systems are operated to permit development to continue where service is otherwise lacking. Developer systems are typically interim utility systems. Expenses are for the construction of water and sewer line infrastructure. Maintenance is typically minimal. Some developer systems are subject to public utility commission (PUC) review, but many are too small to fall under PUC purview. Many developers expect their system to be eventually absorbed into a larger, centralized system—either public-sector or investor-owned systems. The cost to provide an interim system, versus the rewards for development taking place prior to the availability of service from a central system, must be determined by the developer to be worth the investment risk. The goals and objectives of a developer utility owner diverge from the goals and objectives of the rate payer. This occurs because a developer usually enters the utility business only out of necessity for development. Therefore, it is in the best interest of the developer utility owners to build the utility with the capital cost burden placed on the rate payer in terms of high rates, while keeping to a minimum the utility costs (connection fees and impact fees) incurred by the development company. The result is a utility with a high rate base and operating costs and minimal connection fees.

Private-Sector/Investor-Owned Systems

Private-sector utilities are either small, investor-owned utilities that exist within a limited service area, serving a limited area that generally corresponds with some prior development (i.e., they used to be developer systems), or utilities owned by holding companies that provide management expertise and support in operating a series of small, local utilities, typically run with local operators, maintenance personnel, and a local manager. Since operating a utility is very fixed cost intensive, the number of customers connected to a system is extremely important.

At the local level, a private utility system may theoretically have similar operating and maintenance costs as does a public-sector-owned system. However, borrowing money is typically arranged through intracompany loans or taxable bond proceeds on the open market. Both occur at higher interest rates than public offerings (unless industrial development bonds or some similar public financing can be arranged).

Private-sector utility systems are monopolies that have their rates regulated by public-service commissions or other public entities to ensure that the private provider is not taking advantage of its customers. The basis of the rates is often developed from case law that states that the rates must be just, reasonable, and compensatory; they must also balance the right of the public to be served at a reasonable cost and the right of the utility to earn a fair return on the value of the property used to provide the service.

There are three primary factors used in the development of rates: expenses, the rate of return deemed appropriate, and the rate base (depreciated value of infrastructure investments made) upon which the rate of return is calculated. The United States Supreme Court does not permit fair value as the sole basis of rate making. The average investment of utilities during a given test year is used to determine the rate base rather than the value of the property at the end of a test year *(City of Miami v. Florida Public Service Commission).*

Operating expenses include the full gamut of potential expenditures, including salary and wages, maintenance, repair, collection, distribution, extending services, and so on, as would be found in a publicly operated utility, as well as franchise payments made to municipalities, depreciation, and the cost of filing for rate increases. The rate of return is typically based on the following: the returns of other utilities of comparable size and service, the risks assumed at the time of the investment, location, economic conditions, fiscal history, the ability of the utility to borrow money, the necessity of attracting capital and maintaining credit in order that service can be continued, necessary expansions affected, cost of money, dividend interest requirements and related factors associated with attracting capital and maintaining credit, and efficiency of management.

Private, investor-owned systems, if well managed, and absent unusual circumstances, are allowed to earn a specified rate of return on their investments, making investments in these utilities among the lowest-risk investments available in the United States and Canada. From a local-official standpoint, there are both benefits and problems. Control and rates are not issues in which local officials have any say. However, franchise and property taxes can be secured that benefit local governments. Where significant problems exist, the private system may be acquired by local government.

CONSIDERATIONS WHEN FORMULATING WATER SERVICE DELIVERY IN THE PUBLIC SECTOR

In evaluating the delivery of water service, there are a number of factors that public officials must consider, which will in turn provide direction for the water system to pursue. While some of these factors concern the general operation of the water system, others respond to financial or political needs of the community. The operational and financial issues are generally straightforward in their evaluation. Information on revenues, expenditures, budgets, customer costs, and debt obligations can be gathered and compared. Operating factors such as system reliability, system expansion, maintenance needs, and economies of scale are common to all systems. Political factors are harder to predict and evaluate as there is no quantification available. This section is intended to highlight some of the service-delivery issues and provide some conclusions that may fit with local conditions, provide some insight on planning needs, and outline programs that should be developed in every water system regardless of size. Privately owned investor systems present other challenges to local elected officials.

Debt Obligations

Debt affects a local government's financial capacity. Utility system debt can adversely affect a local government's ability to finance other expenses. The most commonly used method to separate utility system needs from the balance of the local government's operation is to set up the utility as an enterprise fund, whereby the revenues of the system are established at a rate to make the water system self-sufficient. The utility system revenues are designed to capture all expenditure obligations, including debt and repair and replacement funding, which are commonly overlooked or underestimated. Debt drives rates, so future bond issues to address deferred maintenance can adversely impact rate payers. More discussion of financing is included in chapter 11.

Control of System

Many local governments want to control the utility system. There are a variety of reasons for this, but the most common are to control or manipulate growth or to use excess water proceeds to balance the general (ad valorem) fund through acquired services or subsidies. The lack of control by the utility system can be politically difficult. The conflicts between local governments and investor-owned utilities in certain areas of the country highlight the differing goals of politically sensitive local governments oriented toward service to constituents who elect them and for-profit investor monopolies that need to make money to satisfy shareholders and state regulatory officials.

Routine Maintenance and Replacement of Aging Infrastructure

Most older utility systems incur significant maintenance costs that are inadequately funded. As infrastructure ages, it is normal for maintenance to increase. Figure 9-1 shows residual value over time. Tracking maintenance costs can only be accomplished by tracking work orders and the investment of time, machines, and parts. Most small water systems do not have tools in place to track maintenance. As a result, repairs are made only after problems have already occurred.

At some point, the annualized cost of replacement can be less than the annualized cost for repairs. At the same time, ongoing maintenance can increase the value and extend the life of assets, though not indefinitely, and never to new condition, as shown in Figure 9-2. As a result, a water system will always be constructing capital projects. Figure 9-3 shows that early in the life of a water system, the investment is for expansion. As the system ages, the costs shift to replacement. Replacement that is delayed is termed *deferred maintenance obligation.*

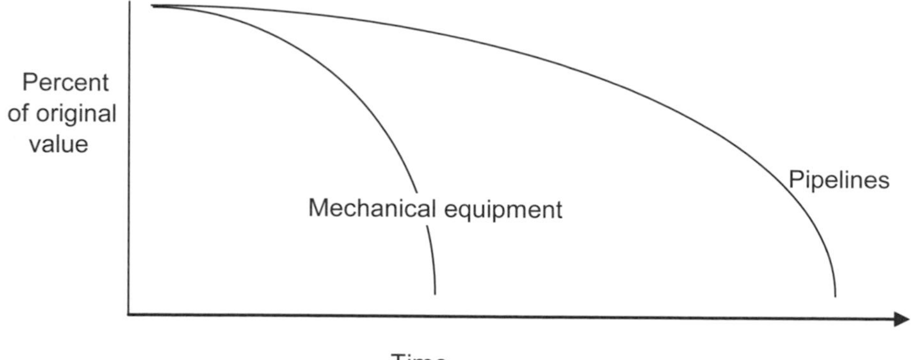

Figure 9-1 Residual value with time

Water systems have traditionally fared poorly in replacing aging infrastructure because rate increases impact customers. USEPA's infrastructure report in 2001 identified more than $250 billion in water and sewer replacement and upgrade needs over the next 20 years. The deferral of replacement of aging infrastructure can lead to system failures and fiscal difficulties in the future. Large amounts of deferred maintenance must be addressed at some point. Delays in addressing the problem add to the cost. Large bond issues are the norm for addressing deferred maintenance obligations, often spurred by significant infrastructure failures. However, borrowing increases rates, leading to long-term negative consequences for future generations of local government officials.

Sufficient repair and replacement funding should be included in the annual budget at a certain level rather than project by project. This will result in reserves that can be used to fund large projects without borrowing. Chapter 11 goes into more depth on this issue.

System Reliability

Directly related to maintaining the system is system reliability. An old, inadequately funded maintenance program increases the likelihood of failure of the water system, despite the best efforts of utility staff. Utility systems need to make service available to all customers at all times, and the regulations are designed with this goal in mind. Public-sector utility operators orient their operations to minimize service disruptions, which have serious political consequences to local elected and appointed officials.

Reliability is an area where public-sector-owned systems and private-investor-owned utilities have differing goals. Private companies assume risk of system failure in exchange for reduced operating costs, because where the private company owns the system, the only risks are bad publicity and fines. Investors are protected. For contract operators operating public-sector systems (privatization efforts), the worst-case scenario is that the private firm is fired, in which case the employees are assigned elsewhere. However, firing a private operator is usually not practical because there are no underlying local government employees

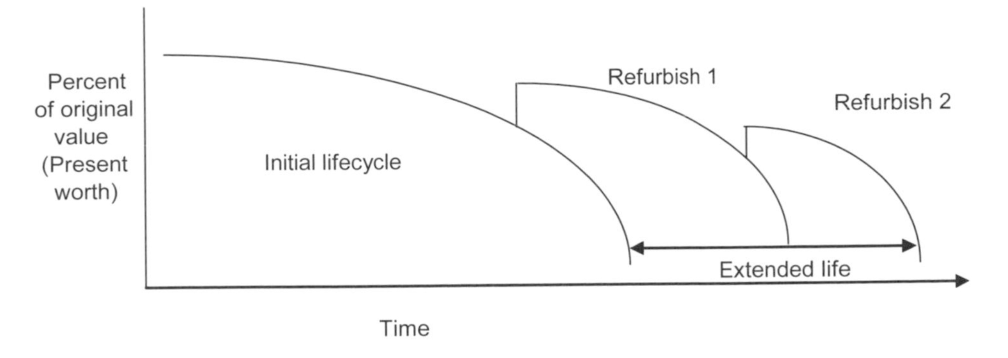

NOTE: Refurbishment may occur later in life before full devaluation takes place. Refurbishment never reaches initial condition. Extension of life of asset decreases with each refurbishment.

Figure 9-2 Value added with rehabilitation

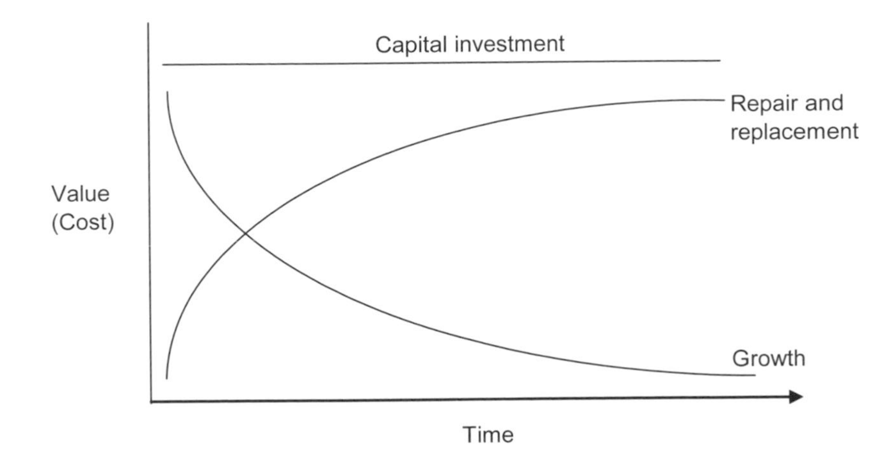

Figure 9-3 Typical utility investment type with time

who are capable of stepping into their place. Hence, there is no real threat of firing until new bids/proposals are undertaken, which often takes six or more months.

Privatization firms use the same philosophy as privately owned systems, which is why their operating costs may be less. Contracts with private firms to operate public-sector-owned water systems transfer most maintenance costs to capital, and the owner is responsible for all capital. Therefore, maintenance is minimal, and the actual costs to the local government may actually be higher.

System Expansion

Who pays for system growth? Courts have indicated that it is appropriate for local governments to have new customers paying enough to fund their costs for operation. This is the reason that impact fees are used in many jurisdictions.

However, marginal cost applications may indicate a need for another approach when new facilities are needed as opposed to using existing, expanded ones. The variation in rates for the new users versus older users is difficult to explain to the public, which is why impact fees and assessments are used. More discussion on impact fees and assessments is included in chapter 11.

Political Issues

Political issues must be addressed in any public-sector operation, less so with privately owned systems. Issues arise with regard to rates, revenue and control issues, and a variety of very site-specific issues. Private systems avoid some of the political issues that public systems deal with unless significant service problems are being experienced.

Fiscal Issues

In most operations, the larger the scale, the less the unit cost. This clearly applies to water systems; large, regional facilities generally have lower operating costs than their smaller neighbors, while providing significantly more expertise and sophisticated operating means. Larger facilities are also generally better operated and comply better with regulations because the cost of needed improvements is spread over a much larger customer base; hence the cost per customer is less.

An area that may impact fiscal issues is the need in local governments to increase general fund revenues. Common general fund revenues include property (ad valorem) taxes and state-shared revenues. State-shared revenues are generally limited in growth from year to year. Ad valorem property tax increases are unpopular. User fees become difficult to impose when there are many of them or when demographics indicate that the costs may affect the public. New revenues from old sources are more popular. "Excess" utility system revenues become a target to supplement other sources as a result. This practice should be avoided (although legitimate costs to the general fund operations should be secured from the utility system, which can be determined via a cost allocation study). The reason to avoid using water funds in the general fund is that this practice reduces available funds for repair and replacement of infrastructure and increases the likelihood of deferred maintenance becoming a problem.

PRIVATIZATION ISSUES

Privatization in the Industry

The American Water Works Association estimates that four out of ten US cities have some form of private involvement within their water systems and another 14 percent of utilities are considering some such privatization options. The most common contracts are for design and construction of new infrastructure and upgrading existing infrastructure (71 percent nationwide), meter reading and billing (33 percent), and distribution of pipe system operation and maintenance (25 percent). Private meter reading is by far the most commonly investigated option (72 percent), followed by billing and collection, then treatment facility maintenance. Several large jurisdictions—Houston, Indianapolis,

Oklahoma City, and Seattle—have significant public partnerships in place to provide infrastructure and to operate the systems.

Current experiences with privatization provide both success stories and failures. The successful projects generally involve awards to firms that minimize costs through introduction of innovations and provision of service. There are few successful ventures where an operator provides a benefit to local governments via the wholesale divestiture of the assets (Blackstone and Hakim 1997).

Private Versus Public Ownership

Some public water systems are owned by private interests, such as a developer, an individual, a partnership, or more commonly, a corporation. From the standpoint of federal and state/provincial drinking water regulations, no distinction is made between private and publicly owned systems in implementing the regulations with regard to requirements for monitoring, operation, and water quality.

This page intentionally blank.

Chapter 10

ADMINISTRATIVE MEASURES

PERSONNEL MEASURES

Technological changes are rapidly making the operation of public water systems more complicated and emphasize the need for greater expertise to deal with ongoing operations and maintenance and plans for the future. Personnel must have the knowledge and abilities to deal with complicated electronics, machinery, and new treatment methods and must be able to understand regulations well enough to maintain compliance. Increased documentation means that administrative staffing needs are greater, as is the need to organize and manage the flow of information among staff, management, and other entities. As a result, it is important that the utility management team has professional credentials and a demonstrated ability to deal with complex utility issues. Access to hydrogeologists, rate experts, engineers, and other scientists has become more important and commonplace.

Elected officials should bear in mind that the safest, most efficient operation of a water system will be achieved by following the principles of employment management, which are

- Professional staff with appropriate education, experience, and licensure
- Salaries and fringe benefits adequate to attract competent professionals
- Selection of new employees based on merit—experience, aptitude, and character
- Viable promotion and compensation policy to keep employees interested in their work

Employment compensation should be commensurate with the education and experience required for the professional management of the utility. The last bullet point involves continuing training programs, for which there are American Water Works Association (AWWA) policies and often state or federal regulations that outline this issue. With regard to career development, AWWA's policy, Employee Training and Career Development, stresses its importance for retention and in ensuring that public safety goals are met:

> It is the policy of the American Water Works Association to:
> - advocate competency-based training and career development within the water profession;
> - develop and offer contemporary, timely training and learning resources for its members;
> - strongly encourage member organizations to formally dedicate time and resources for employee participation in training and career development opportunities, whatever their position or career path.
>
> Competency and continuing skills updates are critical for employees in the water profession, who carry an immense responsibility of guarding public health and safety. Training and skill updates are essential to keep pace with ongoing changes resulting from multiple factors, among them new scientific findings, more stringent regulations, technological updates, increasing customer expectations, aging infrastructure, growing focus on health, dwindling supplies of high-quality source water and increasing costs.
>
> Skill enhancement and career development enhance the water profession's ability to retain qualified, experienced employees and to attract new, highly-capable, enthusiastic individuals.
>
> For all of these reasons, AWWA strongly recommends that member organizations adopt formal policies to endorse and allocate resources to provide competency-based training and career development for employees in all career lines, at all levels and based on equal opportunity. (AWWA 2008b)

The AWWA Employee Compensation policy is similar in its emphasis:

> AWWA strongly recommends that governing boards and water utility managers establish fair and equitable compensation policies that reward the critical elements of protecting the public health and that are competitive with other industries, utilities, and professional services in their service area. It is recognized that public water services contribute directly and indirectly to the general health and economic well-being of the communities they serve and that implemented compensation programs and strategies should be designed to attract, reward, and retain highly qualified managerial, professional, technical, and operating personnel. Therefore, the association urges the adoption of compensation policies and programs to attract and retain employees competent to manage and operate water systems in a manner that will assure safe and satisfactory water service to the consuming public.

Compensation consists not only of direct monetary remuneration for services rendered but also such benefits as medical and insurance coverage, holidays, vacations, educational assistance (including continuing education, skills enhancement, and certification), retirement, and leave for sickness, injuries, and military or jury duty. Benefits, as with salaries, should also be in accordance with general practices of other industries, utilities, and professional services in their service areas.

An equitable employee compensation program should include:

- Equal compensation for work of equivalent responsibility
- Periodic review of the utility's compensation plan and compensation in related industries in both the public and private sector, with periodic compensatory adjustment to maintain a competitive salary base
- Special attention to current conditions and trends in employee benefits because benefits represent a significant portion of the total payroll
- A method of rewarding employees for competent service
- Employee retention plans, including succession planning that offers utility employees work opportunities and special assignments that develop the knowledge, skills, and abilities required in more responsible and/or promotional positions designed to maintain continuity and stability in water utility operations
- Regular review of position descriptions to ensure they reflect operational and technological changes that might impact compensable factors (AWWA 2005a)

Employee Training

Many operations people suffer from a lack of training on new and evolving issues. Both the employees and managers of public water systems should realize the need for proper employee training and should communicate this need to elected bodies. Most water system employees are trained by their supervisor and/or coworkers. A new worker is often assigned to work with an experienced worker to learn the duties to be performed. Unfortunately, many people are not good teachers, and many workers resist sharing their knowledge for fear it will jeopardize their own position. Also, this type of training may perpetuate errors and misinformation if the trainer is not performing the tasks properly. Regulations are in place in many jurisdictions that require certification and/or licenses for many aspects of utility operations. The principal benefits of good employee training are

- Improved morale and interest by employees in their work
- Increased employee productivity, loyalty, and dedication
- Decreased chance of employee errors that cause problems such as damage to equipment or contamination of the water supply
- Improved opportunity for employees to develop into future supervisors

On-the-job training of employees should be a deliberate, conscious effort. Supervisors or experienced employees who are willing to share their knowledge should be designated as trainers and allowed the time necessary to perform this work. Sources of outside training opportunities are also available to water system employees. Training seminars conducted by the state public water supply agency, the state Rural Water Association, AWWA sections, and other water utility organizations include

- Correspondence courses
- Meetings of local water operator groups
- Water utility courses provided by colleges

AWWA has this policy statement on operator training:

> AWWA supports and firmly believes that mandatory continuing education measured by recognized units such as contact hours or continuing education units (CEUs) is essential to the development and sustained competence of all operators. Therefore, the accumulation of these units of education and training should be a requirement of certification renewal.(AWWA 2003)

Operator Certification

Most regulations now require that community public water and sewer systems be under the direction of a certified operator. Some jurisdictions have separate distribution, collection, and treatment plant licenses. Increasing emphasis is expected to be placed on mandatory certification as a means of ensuring that the complex responsibilities of water system operation are under the direction of competent persons. AWWA fully supports mandatory certification efforts as evidenced in this policy statement on operator certification:

> AWWA fully supports mandatory certification of the persons in responsible charge of water treatment and distribution facilities. Further, all operators of water treatment and distribution facilities should be encouraged to gain certification. AWWA believes the model certification program and training coordination procedures such as those developed by the Association of Boards of Certification (ABC) should be followed to ensure the efficient use of state, provincial, and local resources.(AWWA 2003)

Certification programs vary from state to state, but in general, water system operations are divided into several degrees of complexity, ranging from a simple well supply to a full surface water treatment system. The operation of a distribution system is also usually assigned a separate class. An operator must ordinarily achieve certification for each lower class before proceeding to the next class.

Award of certification in each class is usually based on passing a state-administered examination, in addition to having required education credits and meeting an experience requirement of months of service actively working in a water system organization. Reciprocity may exist between jurisdictions that consider they have equally stringent certification programs.

A person can prepare for a certification examination by self-study or by using a correspondence course. Examinations generally emphasize specific points and processes. The best way to prepare for these exams is through a certification preparation course. These courses are held periodically and may be sponsored by the state/province, a water utility organization, or a local college.

Water operators usually must renew their certification every few years. A growing number of states/provinces are requiring operators to have a specified number of CEUs in order to renew certification. The CEUs are awarded for attending seminars and training courses. The continuing education requirement helps ensure that certified operators are keeping abreast of changing rules and technology.

SAFETY

As noted in chapter 2, every utility should designate a safety officer whose duties include ensuring that other employees conform to appropriate safety laws, regulations, and policies. Safety is a major issue for utility systems, since personnel are subjected to trenches, heavy machinery, traffic, chemicals, lightning, electricity, and in the case of meter readers, pets. Losses due to worker injuries can be a serious financial drain on the water system as well as causing a loss of worker time. A safety program will help minimize injuries, thus helping to minimize pain, costs, and disruption of the organization that can occur when someone is injured or dies in an accident.

To minimize job-related injuries, supervisors should be provided appropriate training so that they are knowledgeable about recommended safety practices and can in turn impart this knowledge to their subordinates with the appropriate training materials. Regular safety meetings should be held to remind workers to work safely. AWWA has safety documents that can facilitate these sessions, as do many other organizations. Federal and state agencies periodically review the safety record and program of water systems to ensure compliance. Failure to have a safety program or to document it can lead to fines and liability on the part of the water system. Safety should be emphasized during emergencies when workers are tempted to take shortcuts or work under dangerous conditions to hurry the job. Safe conditions and practices that management should insist on include

- Good housekeeping
- Use of personal protective equipment, including steel-toed shoes, orange safety vests, hard hats, and safety goggles. Breathing apparatuses may be required when entering confined spaces. Confined-space entry rules are common for utilities and should be strictly followed. Each year a number of water and sewer utility workers die from confined-space dangers.
- Prompt attention to all injuries, even slight injuries
- Proper operation of tools and equipment
- Ongoing training

Management has an obligation to furnish employees with proper protective gear. The supervisors have the responsibility in the field to ensure that proper measures are taken. The supervisor should set an example by wearing his protective equipment and insisting that workers make use of and take care of the equipment that is issued to them. It is management's responsibility to deal with supervisors who do not set an example, as the risk for liability will ultimately fall to the utility system.

Workers should be provided with the proper tools to do their work in a safe manner. Many accidents are caused from trying to do a job with the wrong tools (screwdrivers are not chisels or pry bars). It is the responsibility of the supervisor to see that proper tools are available and maintained and that the workers use them correctly.

The safety of the public must also be constantly considered when performing utility work. For instance, if a person is injured falling into an improperly guarded excavation, it not only causes ill will but also places system management and individual employees at risk for a suit for damages. Cones, barricades, detours, flaggers, and other public protection mechanisms should be reviewed prior to beginning work in public areas. Supervisors should receive appropriate training to create work spaces that protect both workers and the public. Regulations permit civil and criminal penalties for violations of the requirements.

Here is AWWA's policy statement on safety:

> The American Water Works Association (AWWA) believes a safe work environment is of paramount importance to protect individuals in the water profession who safeguard their community water supplies.
>
> It is the duty of each utility manager, supervisor, and worker to comply with safety standards and to see that safety is an integral part of every daily work process. Safety must take precedence over shortcuts. As unsafe conditions are discovered, they should be addressed and corrected. Safety practices established by state, provincial, and federal agencies should be regarded as minimum standards by all individuals in the water profession. (AWWA 2004b)

EQUIPMENT PURCHASING

Proper maintenance of production and distribution equipment is essential to the efficient operation of a utility. The sudden breakdown of equipment can cause serious problems, such as loss of system pressure or inadequate treatment. As a result, instructions on each piece of equipment should be maintained in a file, and workers should be allowed time and encouraged to study the information in the file before beginning a maintenance job. If the manuals are not clear or sufficient, many manufacturers are willing to have a representative visit the utility and provide further instruction. When this is not possible, manufacturers will usually provide consultation by phone.

Management attitude toward maintenance is important. There should be a clear policy directing a maintenance program, including specifying work to be done, when it is to be done, and who is to do it, and a system inspection to verify that the maintenance was performed, what was found, and the conclusions drawn from the repair. Maintenance is often a rather thankless job, so managers should make extra effort to reaffirm its need and commend employee efforts.

Personnel performing maintenance work should be specifically designated by supervisors. Not everyone is qualified or experienced to do certain types of maintenance work like electrical repairs, pipe repairs, or pump repairs. To assign responsibility to someone who does not like the work or does not have the aptitude may mean that it will not be done properly.

Tools, spare parts, test instruments, and shop facilities must be made available for workers to properly perform maintenance work. An inventory of the required parts and materials should be on hand at all times so that once work has started, it can progress continuously. Having a proper place to do maintenance work, such as a clean, heated workshop, can also speed the job and ensure that it is performed properly.

Some maintenance work should be planned and scheduled in advance. This is termed *preventive maintenance*. The frequency of preventive maintenance work should generally be in accordance with the manufacturer's recommendation in addition to local experience. Some maintenance jobs may be required weekly to ensure proper equipment operation and prolong the life of equipment. It is often desirable to schedule maintenance jobs by season of the year. Well maintenance is usually scheduled for spring and fall, and treatment plant maintenance is often done when the weather is not suitable for outside work.

Records and reports of repair and maintenance work must be kept to achieve an effective program. Work orders and maintenance systems should record information such as the make, model, serial number, installation date, manufacturer's representative's address, and recommended maintenance schedule. This information should be tied to a work-order system on a computer. Space should also be provided on the work-order form for recording the date and details of work performed and the names of employees performing the work.

EMERGENCY MAINTENANCE

All utility systems can be the victim of any one of a variety of disasters. In general, the problems that can disrupt a water system operation fall into two categories: natural disasters, such as earthquakes, floods, hurricanes, tornadoes, forest fires, landslides, snow and ice storms, and failure of the water source, and human-caused disasters (such as vandalism, explosions, strikes, riots, terrorism, and warfare). Recent terrorist and weather events have demonstrated why all utilities should strive to maintain their systems to minimize the potential for impacts to the system and to understand the system well enough to determine when an incident has happened.

Some potential emergencies can be averted or minimized by advance preparations. For instance, good security at treatment facilities can help reduce vandalism. Beyond the problems that can be prevented, utility system managers must be prepared to act swiftly and efficiently in the event of an emergency.

The first step in developing an emergency plan is to analyze which types of emergencies the water system is most likely to experience and the effect of each. An emergency response plan is then developed to include a list of preparations that can be made in advance and projections of the steps that would be taken in the event of each emergency. Some of the primary steps suggested for a water system to prepare for major emergencies are

- Maintaining utility system records in good order and readily available for use
- Maintaining a contact list of state and local water supply and disaster agencies, equipment suppliers, contractors, key personnel in nearby water systems, and any other person or organization that might be able to provide assistance during an emergency
- Maintaining a good stock of repair parts that might be required in an emergency and a list of where additional parts may be obtained quickly
- Installing standby generators or auxiliary power drives for equipment for use during failure of commercial electric power
- Obtaining tools and equipment that will facilitate emergency repairs and maintaining them in a state of readiness for quick response during an emergency

SECURITY

The events of September 11, 2001, highlighted the vulnerability of targets in the United States to terrorist attacks. While the emphasis has been on terrorism, required vulnerability assessments of water systems have allowed water systems the opportunity to review current security measures, evaluate deficit areas, evaluate vulnerability, and implement contingency plans for such incidents. With a water system there are two areas of concern: contamination of the system by foreign biological or chemical agents, and inability to provide service. Public perception of incidents may have greater impact than a real event (such as the blackout of the Northeast in 2003, in which the first rumor was that the event was a terrorist incident).

Contamination of a large water system is difficult. Surface water supplies may be more vulnerable than groundwater supplies, but both undergo treatment that removes many constituents. Getting access to the distribution system is more difficult and will be localized, so the impact will likely be limited. Accessing treatment and distribution sites requires significant effort by multiple people to cause damage. Chemical releases may be the issue of greatest concern, albeit a localized one.

Disabling the distribution system is easier; without power, the system does not run. If the system has no water, many impacts to local residents can be felt, and firefighting capability will be nonexistent. Fortunately, most utilities have multiple distribution sites with backup power to reduce the likelihood of such incidents. Attacks on water tanks may be more cosmetic but may have a greater effect on the public psyche.

To protect against incidents, the following should be undertaken:

- Coordination with local law enforcement agencies in disaster drills and event planning
- Installation of telemetry at remote sites and interconnection of active sites through telemetry so that people can observe activities in real time
- Alarms and motion sensors at remote sites of significance
- Backup power at all sites where pumping and/or treatment occurs, including raw water supplies
- Elimination of chemicals in gas form
- Adequate records of the system
- Tracking of maintenance activities

The most important thing a water supplier can do is properly maintain the system. This includes maintaining good records of the system, valve exercise and maintenance, fire hydrant maintenance, and replacement of old equipment. Funding repair and replacement will minimize catastrophic failure potential and, in a security incident, minimize the potential for large-scale problems as a result of failure of old equipment. Adequate funding and elimination of deferred maintenance are positive steps.

AWWA has short courses and documents on security issues. The Federal Bureau of Investigation (FBI) and other agencies provide training through local sections of AWWA. Water system managers and operators should take advantage of the training opportunities when available. Many annual conferences provide security training sessions.

PUBLIC RELATIONS

There will be times when failures of the water or sewer system will occur simply because of underground conditions, lightning strikes, severe weather, and other natural occurrences that cannot be predicted. Usually, failures are small—blocked sewer lines, water main breaks that have people out of service for a few hours, or pumps that short out—but ice storms and windstorms put thousands of people out of service every year. Depending on the nature and severity of the failure, there may be an adverse effect on public confidence and perception of the utility, which is the major reason why it is important that there is a continuing public relations and communication effort with the public that maintains the public confidence in the utility prior to a failure. When a failure occurs, the response to the public should be carefully planned and factual.

The following are suggested measures when dealing with the public during failures (which can involve the press and other outlets for publicity). Whether good or bad, in the information age we live in, attention spans are short and perception often matters as much as or more than the facts. Since a failure is perceived as negative, the media and public will expect negative issues. However, keeping these suggestions in mind when speaking to the public or the media helps public relations efforts to maintain the public's confidence in the utility.

- Make sure the speaker is qualified to make the responses; confidence is lost when the speaker is not the one who really understands the issue.
- Keep the answers short and to the point; the press likes sound bites.

- Minimize jargon, which people will believe is only used to confuse the issue.
- Avoid negative words and phrases, numbers, and dire predictions.
- Avoid discussions of risk and cost and avoid making promises, all of which can be misinterpreted.
- Be genuine; body language and the perception of the speaker may be as important as the words said.

This is not an exhaustive list, and no two communities will pursue the same strategies, but planning such a strategy prior to an incident is vital.

It is helpful if there is some planning for responsibilities of staff in the event of a failure. It may be helpful to provide training of specific employees who are identified as having a potential role during a problem. The spokesperson should be calm and instill confidence in the consumer. A system to funnel information to a central point (such as the utility director) should be instituted prior to a failure occurring.

Another issue to keep in mind when dealing with the public is the need to avoid absolutes. Instead, focus should be on the goals of the utility, one reason strategic planning can be useful. The goals when managing an incident are to minimize risk and minimize impacts to the public. Because of the technical and regulatory nature of utilities, plain language and a message delivered straight and to the point are most effective. Speculation, promises, cost, and humor should be avoided. Any presentations should be short—10 to 15 minutes at the most—and reserved only for critical issues if the press is involved; otherwise the message will get lost in the details.

Maintaining good public relations does not have to be a big effort. Some things that inform the public and maintain a good image do not require a lot of work or expense. Some of these efforts include the following:

- Public relations should start at the top, with managers maintaining an open and tolerant attitude toward customers and promoting that attitude with subordinates.
- Meter readers and customer service personnel have by far the greatest exposure to the public of all water system employees. These employees must maintain a good appearance, positive attitude, cheerfulness, and tact in dealing with the public.
- Keeping the public informed of water system work takes extra effort but is appreciated by customers. The person who has the water unexpectedly turned off in the middle of a shower or the home owner who has clothes in the washer when the water turns brown due to main flushing has good reason to be irate. Personal visits to customers are effective if work that will disrupt service or property must be done immediately. Doorknob cards or postcards work well if the work is scheduled for a day or so in the future.
- The use of postcard billing has reduced the ability of water systems to add informational notes to customers along with water bills. Some systems have designed their postcards with enough room to add a short public information message. Others send the postcard bill in an envelope once a year so that public information material can be enclosed.

- Pamphlets on water are available from federal and state agencies and various associations. These can be purchased in quantity or copied for distribution to customers. The pamphlets should include details of upcoming work to be done on the water system and advanced warning and details of rate increases in the local papers and on a Web site.
- Informing children about the local water system can pay dividends. This includes providing information and literature for teachers to use in their classes, arranging for a water system employee to speak to classes, or taking classes on a tour of the water plant. Not only will the children be better informed, but they will take the information presented to them home and project a positive image of the water system to their parents.
- Trucks, tractors, fire hydrants, elevated tanks, and other water system equipment that is seen by the public should be kept clean, well painted, and in good repair. Besides projecting a good image, there is usually a side benefit that employees take better care of equipment that is well maintained.
- Employees should wear uniforms with the name of the employee and the water system noted. Uniforms encourage workers to maintain a neat appearance in spite of having to perform relatively dirty work, as well as identifying the employees as water system employees.

Requirements for water quality and system operation often require systems to make expensive improvements. It is important that these systems begin to get public support for funding as soon as it is known that they are affected by the requirements. It should be emphasized that the requirements are considered necessary to protect the public from dangerous diseases or chemical contamination.

AWWA policies state:

> The American Water Works Association (AWWA) recognizes the critical importance and multiple benefits of clear and timely communication with customers, meaningful involvement by community members and stakeholders, and proactive, frank information-sharing regarding water quality and service. . . . Opportunities for input and involvement are essential to public understanding and acceptance of utility programs and projects. . . . Applied with honesty, openness, and receptivity, these critical aspects of utility planning and operations broaden understanding and support for the growing challenges of providing safe, reliable drinking water while advancing awareness of the value of water. (AWWA 2007)

ACCOUNTING

All public water systems should be operated as businesses, with careful accounting of expenditures, revenues, and property. Proper accounting maintains records of the assets of the operation, the outstanding financial obligations, revenues, and the cost of operation. An interpretation of the accounting records and data is normally presented in periodic reports. These reports may be monthly or quarterly, but the most important one is the year-end report that shows financial standing at the end of the calendar year or fiscal year. Local government records should be audited annually.

Private and investor-owned water systems must follow good accounting practices in order to determine the amount of taxes to be paid and to account to stockholders on profits. An investor-owned utility must also show that the business is being operated efficiently in order to justify to public service commissions the need for adjustment of water rates. More discussion of fiscal policies is found in chapter 11.

OTHER PROGRAMS YOU NEED

Water Accountability

Metering all customers solves two problems: it allows the water system to account for the water that is sold, and it enables the water system to bill people on the basis of their actual usage. A residual benefit is that metered services encourage customers to use water wisely. Estimates of water used for fire flows, hydrant tests, and water main flushing programs cannot be easily quantified but should be estimated. Ultimately, the water system should strive to account for all water and to track the unaccounted-for percentage. The unaccounted-for water is generally system leakage and slow meters, although some theft could be involved from unmetered or unapproved taps. Water systems can obtain revenue from street cleaning, construction uses, and other minor water users through the use of temporary meters.

Except in areas of very porous soil, water leaks tend to come to the surface. Water system and local government employees should be encouraged to report leaks so that prompt repairs can be made. Leaks can damage pavement, create sinkholes, and cause excessive flows in sanitary or storm sewers. Some leaks may be detected by sewer maintenance crews noticing excessive flow in a sewer line. Suspected leaks also may be located by using a listening device that amplifies the noise made by escaping water.

Auditing the water system via leak detection programs is recommended for water systems in need of water conservation measures. Any water system with an excess of 15 percent unaccounted-for water should employ an ongoing leak detection program. Professional leak-survey firms can help water systems start a leak detection program. The rate of return of such programs is usually very short.

Water Conservation

Water availability has become a major concern in the West and in coastal areas of the United States; water supplies are not necessarily abundant where people want to live. Water conservation programs are a method of helping water systems improve the efficiency of the water system and delay the need to find new, more expensive sources. Water systems can also engender significant, positive public relations for themselves through water conservation efforts. AWWA and regulatory agencies support water conservation efforts as means to wisely use and protect the resource.

Experts have compiled large databases on the effectiveness and costs of various conservation measures that incorporate a community's physical and environmental conditions necessary to yield a benefit/cost analysis of the most efficient courses of action to pursue. To identify all of these efforts is the subject of whole books. Analysis of cost/benefit ratios helps a community select appropriate measures, make allowances for budgeting the program, and provide more flexibility in response to changing conditions.

The most effective programs are planned for five- or ten-year periods as part of other planning efforts. Rate experts should be used to develop rates that encourage conservation yet generate the required revenues. Higher-usage categories should be targeted for using conservation measures before low-use categories as they yield the highest cost/benefit ratio to the utility. Because residential customers typically use about half of their water inside the home, the focus is generally on irrigation uses. When conservation programs are undertaken, the utility should always keep in mind the possibility of revenue shortfalls.

Numerous water conservation measures can be employed to reduce water use; some are more beneficial to a specific system than others. In any water conservation measure under evaluation, the savings, costs, and acceptance are critical factors to a cost/benefit analysis. The following measures are commonly adopted water conservation measures:

- A public information program focused on water-efficient landscaping practices and indoor use of water
- Ultra-low-volume plumbing fixture standards that allow no more than 1.6 gallons/flush, 2.5 gallons/minute for showerheads, and 2.0 gallons/minute for faucets, at 80 psi
- Efficient landscaping programs, such as drip irrigation, xeriscape, and water sensors that override sprinkler systems
- Adoption of permanent irrigation restrictions; irrigation should be at night or in the early morning hours
- Institution of a leak detection program, including water auditing procedures, in-field leak detection efforts, and leak repair
- A conservation rate structure (see chapter 11)
- Metering of all customers
- Reuse of wastewater for irrigation purposes
- Pressure control on the potable water system
- Filter backwash water recycling at water treatment plants (where applicable)
- Elimination of potable water use for air conditioning

Where short-term shortages exist, the conservation measures usually consist of restrictions on lawn sprinkling and car washing. Local ordinances may be necessary to enforce conservation efforts.

In those areas of the country where the amount of water readily available is much less than required for unrestricted use and where expansion of water sources cannot keep up with the demand, water conservation measures become essential. Rate mechanisms to penalize high-rate users are effective. Public education programs to instill customer awareness about reducing use and waste are essential.

Fire Protection

Fire insurance rates differ among communities primarily due to the Insurance Services Office (ISO) rating that has been assigned to the community. ISO Commercial Risk Services, Inc. is a nonprofit corporation serving the insurance industry. Communities are classified by the grading schedule on a scale of 1 through 10. Class 1 is a system with a highly rated fire department and water supply facilities and procedures. Class 10 communities have no fire protection within 5 miles (8 km). Forty percent of the rating is based on the water system. As a result, a community with a good fire department but an inadequate water system will probably be assigned a poor rating. Water system rating considerations include ability of the supply works to meet maximum demands; capacity of mains to deliver fire flow; maintenance needs and records; and details of hydrant distribution, pipe size, and maintenance. Details of the water system rating procedures are provided in AWWA *Manual M31, Distribution System Requirements for Fire Protection.*

Cross-Connection Control

AWWA has a policy that states:

> In the exercise of the responsibility to supply potable water to their customers, utilities must implement, administer, and maintain ongoing backflow prevention and cross-connection control programs to protect public water systems from the hazards originating on the premises of their customers and from temporary connections that may impair or alter the water in the public water systems. (AWWA 2005c)

Water distribution systems are under pressure, so the only way a cross connection works is if there is backpressure from the customer or a drop in pressure in the water distribution system due to system damage, breaks, or repairs. The purpose of a cross-connection control/backflow prevention policy is to isolate the potable water supply system from the possibility of backflow, as backpressure and/or backsiphonage may occur because of the existence of cross-connections between potable and nonpotable systems. Both are possible. Backsiphonage can occur when there is reduced pressure or a vacuum is found in the water system that allows a foreign substance to flow into a water system. Backpressure is the situation in which contamination is forced into a potable water system through a connection that has a higher pressure than the water system. Many cases have been recorded of contamination of a public water system as a result of backflow of contamination through a cross-connection, including pathogens, pesticides, herbicides, and nontoxic materials. The conditions under which contamination occurs from a cross-connection are often unusual. Figure 10-1 shows the potential of backsiphonage from a submerged bathtub inlet. Figure 10-2 is a photo of a backflow device.

There are many places in homes, businesses, and industries where unknown cross-connections can be present. Older plumbing fixtures are more likely to

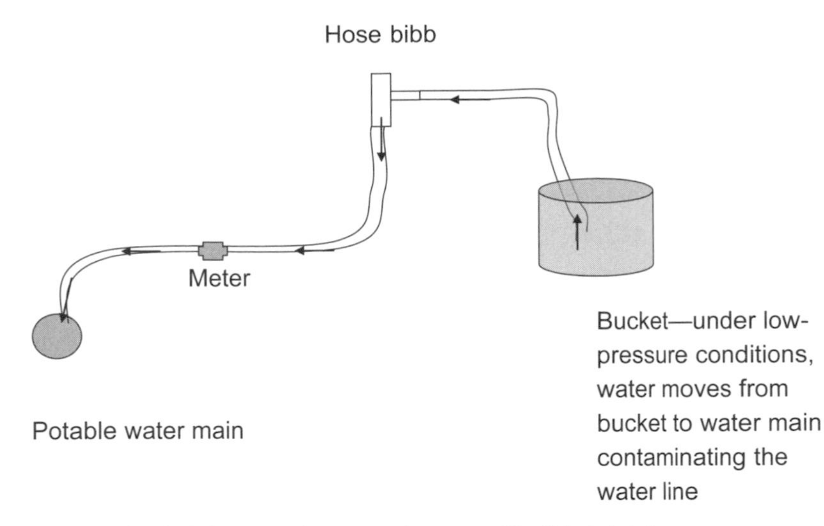

Figure 10-1 Backsiphonage from a submerged bathtub inlet

Figure 10-2 Installed backflow device

possess cross-connections than newer designs. For example, early bathtubs had the fill spout located inside the tub. It is now required that the spout be located above the tub rim to eliminate any possibility of backsiphonage of tub water into the plumbing system.

Most plumbing codes and regulatory agencies require or urge all public water systems to have an active cross-connection control program. Ordinances and other codes should prohibit cross-connections of private systems to the utility's potable water system except when and where approved and appropriate backflow prevention devices are installed, tested, overhauled, and maintained.

In areas with a large number of fire, agricultural, or commercial systems, prevention of cross-connection should be a priority. Model plans are available from AWWA and state/provincial agencies to assist water system managers in enacting appropriate control ordinances, making inspections, and educating customers on cross-connection prevention.

LEAD AND COPPER CONTAMINATION CONTROL

Lead and copper contamination received significant attention in the early 1990s when regulations were promulgated by the US Environmental Protection Agency (USEPA) that required all community and nontransient, noncommunity water systems to monitor for the presence of lead and copper in drinking water. Excessive quantities of lead and copper are associated with a number of health effects. In particular, children who have had excessive exposure to lead are likely to experience a delay in mental and physical development, impaired mental abilities, and behavioral problems.

While there are many sources of lead and copper in the environment, including paint, soils, dust, eating utensils, and gasoline, the first draw of water systems with corrosive water and lead and/or copper piping tended to contain significant quantities of these metals. Rarely is there an appreciable amount of lead or copper in either groundwater or surface water, so it was determined that aggressive waters dissolve minute quantities of these metals from the pipe during low-use periods. The principal sources of lead contamination are lead pipe, lead/tin solder used for joining copper pipe, and brass plumbing fixtures. Copper contamination can be caused by corrosion of copper pipe. The concentration of lead or copper that may be present in drinking water is affected by a number of factors, such as the contact time, the corrosiveness of the water, and the age of the piping.

Most systems completed their lead and copper studies and implemented actions to control corrosion (see chapter 5). However, ongoing monitoring is required for all systems. The testing requires selecting customers who have plumbing systems vulnerable to lead and copper contamination and collecting samples of water that has been in the piping for at least six hours. Many water systems began programs to remove lead-jointed pipe, lead goosenecks, and similar materials from the distribution system. Such programs, because of the age of the piping, may be more beneficial than extensive corrosion control measures. Each utility should evaluate this on an ongoing basis.

Chapter 11

FINANCING WATER SYSTEMS

Coordination of the financial, budget, and operating policies of a utility system allows managers to properly allocate costs to those benefiting from the service, develop pricing strategies that can be clearly explained to the public, and prevent challenges to allocation methodologies. Operations, capital programs, and long-term variability of the utility system operation require financial and facility planning. Multiyear financial forecasts and financial plans are common tools in business and are worthy of consideration by utility systems.

Governmental expenditures are subject to changes in statutes, case law, competitiveness among public entities, the political process, and group decision making. Many public entities rely on projected revenues and subsequent planning expenditures to remain within the revenue projections. Capital items should not be budgeted because they just "fit in." Such a practice can lead to insufficiency in maintenance obligations or the deferral of needed capital expenditures. Since many capital expenditures must be planned years in advance, the consequences of these capital items not coming online at the appropriate time may subject the utility and its customers to excessive maintenance costs, lawsuits, or failures in providing service. Most progressive utilities today project expenditures (including long-term capital allocations), a practice which causes revenue needs to become clearer from year to year. Projected shortfalls can be planned for ahead of time, and capital expenditures can be scheduled and completed at the necessary time. Rate increases can be phased in to minimize impacts on customers when expenditures are projected.

FORECASTING REVENUE AND EXPENDITURE NEEDS

Forecasting revenues and expenditures should be long range in nature to be of any use to the utility. The programming and forecasting of both revenues and expenditures for at least five years is basic to the conduct of any activity or service of a government entity and essential for sound financial planning. In forecasting changes in services, personnel, plant, and materials needs must be considered. The plan should be adjusted and updated annually as a part of the budget process. The forecast will then be of significant benefit in preparing the annual budget (some enterprise funds may require longer-range forecasts due to the high cost of the improvements).

Capital programming should only be a portion of the forecasts. While capital items are important, operating requirements may often be larger than year-to-year expenditures. Long-term changes in service demands, revenues, socioeconomic factors, and changes in state/provincial and federal revenue programs should be taken into account. The benefits of these projections are numerous because they

- Facilitate the determination of priorities for services
- Promote adherence to orderly maintenance schedules to avoid significant deferred maintenance accumulations
- Cause a review of user-charge rates and other revenues for adequacy
- Provide a window of opportunity for planning bond issues and debt service obligations
- Encourage proper timing of major expenditures to reduce budget impacts
- And perhaps most importantly, allow for identification of opportunities to address fiscal gaps and revenue shortfalls

COST ALLOCATIONS

The American Water Works Association (AWWA; 2005b) believes "the public can best be provided water service by self-sustained enterprises adequately financed with rates based on sound accounting, engineering, financial, and economic principles." This is in keeping with the recommendations of most professionals and case law. Water and sewer utilities are generally designated as enterprise funds, separate from the general ad valorem tax funds of local governments. This is done for two reasons: the customer base and tax base may not be the same, and the cost of service for the utility system should be related to the service charge.

Once the revenues and expenditures for the system have been forecasted, the next issue to address is the appropriate allocation of expenditures to revenue sources and identification of any restrictions that may apply. Case law is based on the legal precedent that a utility may charge for its service and products it provides to its customers, but the basis for the charge must be reasonably related to the cost of service or the product. While case law varies among states, it is useful to keep the idea of fairness in cost allocation in mind. The utility system's rates must be reasonable and nondiscriminatory, although different user classes can be charged differently, provided a valid rationale exists for the difference. While public-sector-owned utility systems are allowed by case law to make a profit, this profit has been held to be valid at levels less than 30 percent of total operations. As a result, *municipal governments must exercise caution when transferring unspent water and sewer fees to the general ad valorem tax fund to subsidize general fund operations.*

Different user classes may be charged different rates if the rates can be justified. For example, residential customers can be charged differently than industrial or commercial customers. Also, users of a new water treatment facility can be charged differently than those using an older facility if there is an obvious separation between two treatment facilities being used by the residential customers. However, impact fees and other revenue collection methods are perhaps more appropriate mechanisms to absorb this difference and provide consistent rates across all user classes.

Consideration should be given to funding repair and replacement funds. Repair and replacement funds are collected from existing customers to repair and/or replace existing infrastructure at the requisite time. Periodic service charges, broken down between availability and volumetric portions, are utilized to collect operation and debt service from customers receiving the service. Connection charges and a variety of miscellaneous fees are utilized for particular types of activities, including but not limited to the cost of making taps, rereading meters, meter testing, backflow tests, and contractor fire hydrant tests. Each of these opportunities will be discussed in this chapter.

The budgeting process should include the costs for debt first, repair and replacement dollars second, and operations third, thereby establishing the total need for revenues. Adjustments can then be made, preferably to the operations budget and not to the capital or repair and replacement budgets. Periodic rate increases, generally reflecting inflation, should be expected and perhaps planned in multiyear rate resolutions.

Availability Charges

The fixed-fee portion of a service charge is collected from every customer regardless of whether or not there is any water usage at the address. This practice is intended to allow a utility system to bill customers where service is available, because there is a cost for just having the service available to the customer's property. One obvious and consistent charge encountered is that of meter reading and sending out the water bills. As a result, meter reading costs should always be included in the fixed portion of the bill. Likewise, debt service continues to occur whether or not the customer uses the system, so revenues to cover debt are typically included in the availability charge. Debt is the highest priority in the budgeting process because most debt instruments require the utility to take whatever actions are necessary to repay the debt. Including debt costs in the availability charge of the utility bill provides a safeguard in event there is catastrophic facility damage due to storms or other natural disasters; the availability charges continue to accumulate on the system to enable the utility to pay its debt, even though the service is not being used.

The problem with this rationale is that in areas with large numbers of people on fixed incomes or with limited economic ability, or where conservation is a goal, the associated higher, fixed fee may not be politically palatable. However, all these issues can be addressed alternatively without compromising a system's stability or financial objectives through guaranteed minimum base amounts of water or other methods developed by creative rate consultants. Two methods are used for creating the availability charge: an equivalent residential unit (ERU) system and a meter-sizing system. A rational argument can be made for each, although each has its drawbacks.

Meter-Sizing Method of Allocation. The argument for using a meter-size methodology is that the benefits to all users with similar-size meters is the same since a given meter size has an average and maximum flow capacity and each user has an equal ability to draw water from the system. Under this methodology, each similarly sized meter should be billed the same fixed amount because the customers' access to water is the same, assuming it is within the design parameters of the meter (AWWA has standards on this; see Table 8-2). The fact that they may not use the full available supply is a matter of choice and in many cases is dictated by economic and conservation attitudes.

A typical, single-family residential unit utilizes a ⅝ × ¾-inch meter. All other meter sizes can be calculated to be a multiple of this size (see Table 8-2). Once all the equivalent meter sizes are defined, the debt and meter reading costs can be apportioned appropriately, based on the ⅝ × ¾-inch meter size. However, there are two precautions to consider when using this system. First, the existing meters must be sized correctly. This is often a problem with existing customers, who will resist making the upgrades to their meter and system (and paying increased amounts for service). In addition, residential users are generally subject to peaks (morning and evening) and virtually no use during the day (a diurnal curve). As a result, when sizing multifamily unit meters, the peaking factor must be used, not the average flow, which may dramatically increase the meter size.

ERU Method of Allocation. The ERU method resolves problems with sizing multifamily residential meters noted with the meter-size method because it accounts for concurrent peak usage. It also addresses historical undersizing of meters that often has occurred (causing utilities to undercollect from trailer parks, condominiums, and apartment buildings as is common when using the meter-size system) where the political will does not permit the forced upgrading of substandard systems. The ERU methodology assumes some control is exercised over the demands imposed by the ERU, whether it is defined as a typical residential single-family unit, an average gallonage use (often between 250 and 350 gallons per day for water), or some other system. This system is generally used in residential communities, and often condominiums, apartments, and mobile homes are deemed to be a partial ERU (0.7 is common).

The concern in using the ERU methodology is the method of conversion of nonresidential or dissimilar residential units such as travel trailers and extensively landscaped homes to the ERU basis when their usage can vary widely. Table 8-2 can be used, but there is no clear demarcation that justifies a given commercial use as a multiple of an ERU. As a result, most utilities pursue a hybrid approach of the two systems—ERUs for all residential users and meter sizing for all other uses. While the line of demarcation between residential and nonresidential is not always clear, such as in the case of RV parks and retirement homes, the hybrid approach mitigates concerns when using both systems.

In calculating the availability charge, the number of meters read should be considered. When there is a single large meter servicing many ERUs, the debt associated with each ERU, but not the meter reading costs, should be assigned under the ERU approach. As a result, when calculating availability charges, there should be two considerations, one for the cost of reading the meter and one for the cost of debt. A large meter should have only one reading charge.

Volumetric Rates

With the exception of debt, meter reading, and billing costs, all other costs of the utility's operation are usually assigned to the volumetric portion of the bill. Most of the remaining costs are generally dependent on the amount of water or wastewater that is treated and pumped. The larger the utility, the greater the quantity of water being treated, the larger the piping system to maintain, and therefore, the more staffing is needed. Economy-of-scale studies demonstrate that larger facilities have fewer operators per gallon of water treated and lesser costs, all other factors being equal (one argument for regionalization is cost savings).

Included in the volumetric charge of a utility system are the planned expenditures in the operating budget, including but not limited to

- Water and wastewater treatment
- Pumping and transmission
- Pipeline maintenance

Nonoperating and support service costs include administrative costs, which would include

- Purchasing of materials
- Legal fees
- Insurance
- Budgeting and accounting
- Human resources
- Engineering
- Public information
- Technical information
- Dissemination of operational needs
- General management
- Public relations

Allocation of these nonoperating costs should be apportioned proportionately over the operating budgets or personnel.

Water Rate Structures

The cost of supplying utility services has increased significantly over the period from 1980 to 2000. This increase occurred for many reasons, including:

- Stricter regulations established through passage of water quality regulations
- The need for many utilities to develop increasingly more costly water supplies
- The rapid economic development in certain areas of North America

- The onset of droughts in some areas
- Higher cost for development of new water sources throughout the country

Traditionally, utilities have used water pricing as a means to recover costs, by charging users of a specific type in accordance with the cost of serving that type of user: this water pricing practice is both effective and equitable. However, pricing can also work to reduce demand by providing an incentive for customers to manage water use more carefully. The customer's sensitivity to pricing is enhanced by rate structures that increase the cost of water at a constant or increasing rate as usage increases. Since there is significant flexibility in structuring rates to achieve this objective, it should be the goal of the utility to select the rate structure that achieves cost-of-service principles and meets the community's needs. Among the considerations in developing a rate structure are

- Financial sufficiency—generating sufficient revenues to recover operating and capital costs
- Conservation—encouraging customers to make efficient use of scarce water resources
- Equity—charging customers or customer classes in proportion to the costs of providing service to customer groups
- Ease of implementation—having the capability to implement the rate structure efficiently without incurring unreasonable costs associated with reprogramming, procedures modification, and redesigning of billing forms
- Compliance with appropriate legal authorities—being consistent with existing local, state, and federal ordinances, laws, and regulations
- Effect on customer classes—minimizing negative financial effects on utility customers
- Long-term rate stability—producing rates that are reasonably constant from year to year, (i.e., the methodology does not produce rates that fluctuate widely from one period to another)

It is important to project operating and capital costs over an extended period of time so that fluctuations in potential rates can be evaluated. It is also essential that revenue requirements are sufficient to provide adequate facilities, to allow proper replacement and maintenance, and to ensure that the utility is operating on a self-sustaining basis. After costs have been allocated to user classes, it is then necessary to design a rate structure of appropriate charges to customers. Precedent and reasoning used in developing water rates must be sound, so that conclusions reached can withstand scrutiny in the event of litigation. The development of water rates to achieve political and social objectives should be minimized in the rate design to avoid the appearance of discrimination.

When selecting a rate design, a utility should consider its operations and economic environment, the community's objectives, and its customers. To achieve its objectives, a utility should adopt a comprehensive process for evaluating the appropriateness of alternative rate structures. Factors to consider in revising existing rate schedules are

- Effect on customer groups or individual customers
- Compliance with local, state, and federal laws and regulations
- General public reaction to changes in rates
- Impact of shifts in the cost burdens from a group of customers that has been overcharged to a group that has been subsidized under the existing rates
- Reluctance to depart from the existing rate form because of tradition
- Pressure from special interest groups
- Ease of implementation
- Equity
- Simplicity and understandability by customers
- Water conservation

The development of a rate schedule to meet costs of service will have to take into consideration local practices and conditions and should be in the best interests of both the community and the utility.

The application of rates to maintain a sound utility system is a policy decision. This decision can be accomplished by reviewing past practices, considering present and future effects on customers, and providing rate modifications that will accomplish the necessary results with the least controversy. It may be necessary in the application of a developed rate structure, such as one for conservation rates, to spread the modifications over an extended period of time to accomplish the desired results. Designing rate structures that conserve water is a requirement of consumptive use permits granted by some regulatory agencies and may become an integral part of the water supply planning process. The following paragraphs outline the more common rate structures and their goals, benefits, and problems. Most of these rate structures have more applicability to potable water systems than wastewater, due to the need to conserve potable water.

Declining Block Rates. For many years, a single schedule of declining block rates applicable to all customer classes was the predominant water rate structure in the United States. The declining block rate provides a means of recovering costs from the customer classes under a single rate schedule, recognizing the different water demands and costs associated with each customer class. Under this rate methodology, economies of scale are recognized since the price per unit declines as the water customer consumes more water. This methodology provides no incentive to conserve water; in fact, it encourages just the opposite. Recent rate surveys indicate that the traditional declining block rate schedule applicable to all customer classes, once popular in industrial cities, may no longer be the primary method of charging customers as a result of shifts in industrial uses and the need for conservation of water resources in many parts of North America.

Uniform Volumetric Rates. A uniform volumetric rate is one in which all water (and/or sewer) use is charged at the same volumetric rate to all customers. This is the rate structure typically utilized with sewer systems. Sewer rates are normally based on water meter readings (although many systems cap the sewer usage at some point); sewer is not metered for the majority of users as metering is difficult to accomplish on gravity-based flow regimes. With uniform volumetric rates, there is no change in the unit rate for water/sewer used during the billing period. The uniform rate may encourage water conservation by eliminating the lower-priced water rates under a declining block rate structure. The uniform rate may also be perceived in some communities as being more equitable because all customers pay the same rate, regardless of water use volumes. The amount of conservation to be expected when using a uniform rate structure is minimal.

Inverted Block Rates. Inverted block rates are the opposite of the declining block rates. Under this alternative, rates increase for progressively larger volumes of water use. As a result, larger-volume customers pay a progressively higher average rate for increased water use. The reasons for using an inverted block rate structure are to offer financial incentives for reducing water use and to include a price for the additional capital facilities necessary in order to meet peak demands (i.e., a high correlation of peak demand with water use by large users).

Some utilities apply one inverted block rate system that is applicable to all customers. Other utilities adopt separate inverted block rates for different customer classes. For example, residential customer classes may be charged at a higher unit rate for water use in excess of an established amount per billing period (although if a sewer cap is used, the block rate should be higher than the total of water plus sewer at the cap limit). Typically, the higher-priced block(s) are established at amounts that would typically include nonsanitary usage such as irrigation, car washing, and filling swimming pools. Under this alternative, these uses are charged at a higher rate in order to discourage excessive usage. A variation of the inverted block rate structure could include a mechanism whereby all water use is charged at higher rates if usage exceeds a certain preset amount. This element can provide greater conservation incentives to the customer since excess water use will cause all purchases to be charged at higher rates, not only the usage in the higher blocks.

The ability of a price increase to affect consumption is termed price elasticity. Price elasticity of demand is a measure of the relative influence that a change in price of a commodity (water) has on the demand for that commodity. Water savings due to this type of increase in cost are dependent upon the price elasticity of the demand. Price elasticity can affect long-term savings of inverted block methodologies. Several variables have significant effects on price elasticity; for example, elasticity varies with the type of customer, the income of the customer, and the amount of water used. Although in most cases a price increase will reduce consumption, the magnitude of the reduction is dependent upon how the customers react to the financial increase. Many utility systems have adopted inverted rate structures for conservation purposes. The use of inverted block rate structures appears to yield some overall reduction in water use, at least in the first few years.

Off-Peak Rates. Off-peak rates apply to water service used during periods when the utility is not providing water service at its daily- or hourly peak rates of flow, in much the same manner that power companies rate electricity service.

This rate reflects lower utility costs during these off-peak periods. Off-peak rates may encourage customers to take larger portions of their water needs during off-peak periods. By shifting some demand to off-peak periods, the existing facilities are less likely to be overtaxed during peak periods, and the subsequent need to construct new facilities to meet load growth may be reduced. This rate structure is used widely for electric utilities but is nearly impossible to implement for water systems.

Seasonal Rates. A variation of off-peak rates is seasonal rates. Seasonal pricing to affect peak use is probably the most effective and equitable way of managing demand. Outdoor usage, which is responsible for most peaks, provides an opportunity for generating the greatest reduction with the least burdensome pricing strategy. Seasonal rates establish a higher rate for water use during the utility's peak demand season, reflecting limitations in supply during those periods. Seasonal rate structures indicate to the consumer the importance of efficient use of resources. Such rates are becoming a popular and effective rate structure in areas where seasonal peak uses are high (for example, in beach communities where most of the water is used for irrigation and tourism).

Several variations are possible, but each charges higher rates for use during the peak seasons than during nonpeak seasons. Water use during peak-demand months can be either surcharged or simply charged at higher rates. Alternatively, all water uses that exceed predetermined usage levels during base months (such as winter-period usage) can be charged at higher rates, with the established base usage charged at a lower rate.

Flat Rates. A flat rate is one in which all water use is charged at the same amount regardless of the amount of service received. This rate structure may still be utilized with some sewer systems. Meters are not required. The practice encourages (even rewards) waste. Most utilities with flat rates have changed this system to accommodate requirements in grants or loans received for past construction. The rate system is not equitable except when all users use the same amount of water—condominiums could be an example. This rate structure is not favored by regulatory agencies and is generally viewed as unfair to some sectors of customers.

Schedules by Customer Class. Another option is to establish a separate rate structure or schedule of charges for each customer category served by the utility. However, a rate structure applicable to all classes of customers cannot reflect the cost of service for any particular customer group. By establishing rates by class of user, there is a more direct recovery of cost from each customer category. Since the rates can better reflect cost differences among the various classes, the customers in each user class are made aware of the cost of each unit of water consumed. The major difficulty in establishing a rate schedule by class is identification of the various classes and assignment of each customer to the appropriate category: nearly everyone can make an argument that they belong to a special user class.

Impact of Alternative Rates. Many of the alternative rate structures discussed have not yet been adopted extensively across the United States. A utility using such designs could find itself with rate inconsistencies, competitive problems with other utilities, or court challenges by customers over fairness of the rate structure. For example, some alternative rate structures could affect larger water users heavily and discourage economic development within a particular area. As a result, the utility adopting an alternative rate structure should be aware of the pricing approaches of similar, adjacent utilities.

The rate structure implemented by the utility must meet its needs and policies but may be significantly different than the utility's current rate structure. These differences can have major impacts on the utility and its customers. An implementation plan should be developed that recognizes and addresses these effects. The implementation plan should include provisions for phasing in and testing the impacts of the proposed alternative prior to its full implementation.

Any increase in periodic availability or volumetric rates will have a corresponding percentage decrease in actual water usage. In performing an analysis of the expected revenues, lower usage should be taken into account, regardless of the rate structure used, as each has its benefits, disadvantages, and a different economic response to differing price elasticity.

Any time there is a significant change in required revenues, the effects on the utility customers must be gauged, which may require an updated cost of service to identify areas where cost savings can be realized. When these changes affect different customer categories, it may be beneficial to review a rate structure that will charge the differing customer groups accordingly, so that one class is not unduly subsidizing another.

A comparison of typical periodic bills is the most common way of determining rates and their effects on the consumers. However, in doing this evaluation, a variety of meter sizes and water-use ranges should be utilized to provide a broad-spectrum picture. Too often, rate consultants compare only average bills, failing to look at large users and small users who may be affected in drastically different ways.

IMPACT FEES

Impact fees, charges imposed against new developments or connections to provide the cost of capital facilities made necessary by that growth, may be a United States phenomenon. Generally the *capital facilities* are deemed to be treatment facilities, wells and surface intakes, and regional transmission systems (infrastructure installed in front of individual houses is not an appropriate use of impact fees because it serves no regional value). The use of impact fees is based on case law derived from *City of Dunedin v. Contractors and Builders Association of Pinellas County, Florida,* where a utility's "water and sewer facilities would be adequate to serve its present inhabitants were it not for drastic growth, it seems unfair to make the existing inhabitants pay for new systems when they have already been paying for the old ones." This case is the basis for much of the current impact fee case law in the United States.

Impact fees have been extensively litigated in Florida, less so in other locales. Florida case law is cited in impact fee cases throughout the United States, and its basic tenets are upheld. As developed under this case law, impact fees must meet the "dual rational nexus test." The first prong of this test requires that there be a reasonable connection between the anticipated need for additional facilities and anticipated growth *(Hollywood, Inc. v. Broward County)*. The second prong requires that there be a reasonable connection between the expenditure of impact fee revenues and the benefits derived by new connections *(Hollywood, Inc. v. Broward County)*. In addition, case law requires that these fees be just and equitable. As a result, a profit cannot be earned on impact fees; they must be related to the actual cost of providing the service as defined in the second prong of the dual rational nexus test.

Utilities have instituted impact fees as a method of generating revenue from new customers to finance major facility construction made necessary by the addition of those new customers. To meet the dual rational nexus test, these charges are typically based on the incremental or marginal costs of providing the service, an average cost to provide an incremental portion, or an estimate of the cost of the construction to be provided. Because facility planning timelines may be extensive, and because of the geographical variance in growth demands, a multiyear estimate is used to forecast needed expenditures and proper impact fee amounts.

The impetus behind impact fees is the sentiment to have growth pay for growth. The magnitude of impact fees varies throughout the country, depending on the municipality or the utility's desire to encourage growth. For utilities in Florida, impact fees gained considerable favor after passage of the 1985 Growth Management Act, which required localities to have capital infrastructure (including water and sewer service) available at the time development actually occurs. These large facilities are generally financed with both bonds and impact fees. Other states have looked at this model.

In establishing impact fees for water and sewer services, the findings typically made by governing bodies are as follows:

- The land regulations and policies require owners of land to connect to regional facilities when they become available.
- Those placing additional demands on the system from growth should contribute their fair share to the cost of improvements and additions to the regional system.
- These contributions are an integral and vital element of the regulatory and growth management plan.
- Capital improvement planning is an evolving process defined by a level of service adopted by the governing body.
- Impact fees will protect the interests of the citizens currently served or intended to be served by the utility system, which enhances the health, safety, and general welfare of the residents and landowners within the utility's service area.
- Impact fees are an important source of revenue.
- The deficiencies that exist between the existing system and the adopted level of service cannot be funded through impact fees.

All properties that are connecting to the utility system are subject to payment of impact fees at the time of connection to the system (generally when building permits are applied for), in addition to any costs for installation of subdivision infrastructure (normally both are paid as a part of new lot costs).

During growth cycles, the amount in impact fees collected can be considerable. However, since they are tied to growth, significant fluctuations may occur from year to year, based on local and national economic conditions. As a result, the revenues are not always predictable, making pledges toward debt service of these funds difficult without supplemental revenue pledges. High impact fees may discourage the growth for which the impact fees are intended to pay. In areas that are trying to grow in order to continue growth of the tax base and services, high impact fees are a problem. However, having a subsidy by current rate payers to encourage growth may be equally unsatisfactory. In other areas where growth is too rapid, impact fees charged at the full (not subsidized) cost of providing the facilities may help to control growth.

In determining the value of an impact fee, an important consideration for any defense in the event of a challenge is that the impact fee should reflect the incremental costs to provide the treatment and transmission capacity to the consumer. As such, the present value of any debt service amounts that would be paid during the life of a customer being connected to this system on a current debt should be deducted from the value of the impact fee. Customer impact fees should be determined by a similar methodology. For commercial users, meter sizing is an appropriate method to calculate impact fees since meter size represents the average and maximum available water supply at that address. Since the use at the property may change over time, unless the meters are changed, there is a certain maximum amount of water that can be utilized and remain within the design parameters of the meter. This is the rational nexus for establishing impact-fee rates using meter size, although proper sizing of the meter must be assured.

In developing the appropriate funding levels for impact fees, the options for funding the capital projects anticipated to meet future demands must be established. This includes separating funding sources for repair and replacement projects, projects to address deficiencies in the current system, and future growth.

SPECIAL PROJECTS AND ASSESSMENTS

There are a number of instances where impact fees do not apply since facilities have no regional benefits. Such facilities would include

- Small gravity sewer lines
- Local water lines
- Neighborhood pump stations and attendant force mains
- Facilities typically installed and dedicated to the utility at the time of construction of subdivisions or developments by developers, or retrofits
- Connections to the utility system

Any improvements serving a limited geographical area are generally termed subdivision infrastructure. This would include the installation of water or wastewater service where it is currently not available, or replacement of outdated or older infrastructure on the system. In many cases, the appropriate way to fund these improvements is through assessments. In addition, residents may request that water or wastewater service be replaced or supplied to their neighborhood and may petition the utility's governing board to undertake the project. Assessments are an appropriate means to accomplish these projects.

Assessments are collected to meet special benefits for a sector of the population, and they must represent a fair and reasonable portion of the cost to each of the projects subject to the improvements and the assessment (the assessment version of the rational nexus test). Payment of the assessment bill may be enforced through a lien against the property, most easily accomplished by placement of the assessment on the property tax bill so that failure to pay the property tax bill and assessment (which cannot be separated) will cause the tax collector to pursue liens on the property. While a detailed discussion of assessments is beyond the scope of this document, there are fairly strict requirements established for assessments in state statutes.

CONNECTIONS AND MISCELLANEOUS CHARGES

The category "connection and miscellaneous charges" pertains to a variety of services that are provided to specific people as desired or required. For instance, when someone wants to connect to the system there is a charge to install the service tap, to turn the meter on, and so on. Each of these should be listed on an individual bill to the customer in the amount associated with the actual cost to provide the service (rarely, this is the periodic user charge bill). An analysis should be undertaken to determine appropriate charges for these services, including personnel, overhead, vehicles, and so on. Such things as connections, turn-ons, turn-offs, meter rereads at the customer's request, backflow tests, meter tests, meter change-outs, meter installations, repair of damage caused by the customer, and water audits can all have fees assigned to them. The rational nexus for establishing these costs is that the only person benefiting from these services is the customer receiving the service.

Appropriate allocation and charging for these services can be a significant source of revenue for a utility system in a given year. When concern exists about too many charges for too many services, it is generally because of political resistance or computer limitations in tracking billing. A good computer system will alleviate many of the problems. Any of these services can be included within the general rate for monthly service.

BULK USER RATES

Bulk service is the delivery of water to a neighboring utility or other user on a wholesale basis, bypassing much of the distribution and collection system costs of the wholesaling utility. As a result, the cost to maintain the extensive piping systems and the connections to individual customers is borne by the bulk user, not the wholesaler, so the cost allocations may differ significantly from those for retail customers.

In trying to determine the appropriate rates for bulk users, it is recommended that the availability and volumetric charge system be implemented. The debt service portion of the cost must always be recovered, often on the basis of capacity reserved. This is best done through collection of an availability charge. Secondly, the treatment and transmission costs need to be determined. While cost of treatment is readily available from treatment plant budgets, the cost of transmission is not so easily determined, because transmission cost is often included as part of water distribution or sewage collection budgets, which include both the transmission system and local subdivision infrastructure. As a result, when allocating these costs, careful consideration must be given to maintenance costs per mile for large-diameter pipelines and force mains and the cost to maintain large lift stations. Where maintenance records exist, it may be beneficial to allocate the large pump station costs or tank costs based on past experience. Guarantees of water quality in the bulk user's system after it is passed through the meters and any growth needs or pretreatment limitations should not be included.

A rational basis for demonstrating the costs of operations and maintenance to the bulk users will alleviate a great many potential problems. In trying to determine the formula for a bulk user, establishing a per-thousand-gallon charge along with the availability charge treats the bulk user like a regular retail customer and alleviates the need to do follow-up corrections or adjustments on an annual basis. The cost per thousand gallons to the bulk user usually will be significantly less than that to the retail user.

In many cases, a portion of the bulk user agreement should include a payment of impact fees, collected by the bulk user and paid directly to the water utility. This alleviates the need for calculations for reserved capacity in the availability portion of the bill. However, when reserved treatment plant and/or transmission capacity is desired, a cost to provide and reserve that capacity should be included in the bulk user rate, ideally as part of the availability charge.

OTHER BILLING CONCERNS

It is relatively common for municipally owned water systems to include sewage, garbage collection, or other charges along with the water bill. Sewage charges are structured in various ways that are often in proportion to water use. If a municipality uses the water bill to collect for other services, they should be clearly identified as separate items on the bill because case law indicates that customers cannot have their water turned off for failure to pay nonrelated services. Sewer and water are related. Garbage and stormwater are not. In addition to causing potential refusals to pay, billing other items on the water bill confuses the customer about the cost of water—most customers do not differentiate the charges. As a result, if practical, a separate method should be used for billing services other than water and sewer.

Most water systems have gradually changed from sending bills in an envelope to postcard bills. This saves the cost of an envelope and the labor of stuffing envelopes, and it requires less postage.

FUNDING CAPITAL IMPROVEMENTS

Whole texts have been written about the funding mechanisms for capital infrastructure. This document only presents a brief description of the major funding mechanisms for municipal systems. Appendix A outlines other sources.

Grants

Grants or other awards of funds that do not have to be repaid are sometimes made to water systems by various government agencies. The availability of these funds for utility systems is minimal except to economically distressed, small communities.

Bonds

Bonds are frequently issued by water systems to acquire land, replace outdated or failing equipment and facilities, and expand the system. Bonds generally lead to rate increases, so pay-as-you-go systems are preferred when practical. The bonds provide large sums of money when needed and permit repayment at a relatively uniform level over a period of 20 to 30 years. Interest rates are based on the creditworthiness of the utility system. Small systems in poor financial condition will have difficulty attracting buyers of their bonds at reasonable interest rates. Large systems routinely sell bonds.

There are many types of bonds that may be issued, but the two that are most common are general obligation bonds and revenue bonds.

General Obligation Bonds. General obligation (GO) bonds pledge the full faith and credit (i.e., taxing power) of the municipality against the bonds, despite the fact that the bonds are usually paid off mostly or entirely from utility revenues. The advantage of a GO bond issue for the bondholder is that the added security of having both water and sewer revenues and potential tax income to meet the obligation may secure a more favorable interest rate. The disadvantage of GO bonds is that the bond issue becomes part of the municipal debt and will be included in determining the remaining bonding capacity of the municipality. This obligation can seriously restrict the ability of a small municipality to issue GO bonds for road construction or buildings. A GO bond issue must also be approved by the voters, which is often quite a challenge.

Revenue Bonds. Revenue bond issues pledge the revenues of the water and sewer systems to pay the interest and redeem the bonds when due. Revenue bonds can usually be issued much more quickly than GO bonds, because they do not require voter approval. However, because the revenues of the system are the only pledge (which is weaker than full faith and credit), the interest rates are slightly higher and reserve funds are required. Bond attorneys can develop creative revenue bond issues for specific circumstances.

In both cases, when bonds are issued before the project is bid or completed, additional revenue may be needed that cannot be secured from additional bonds. This is why any small local government looks to state revolving fund loan programs or other loan options.

Low-interest loans are sometimes available to publicly owned utility systems from state/provincial or federal agencies, under varying circumstances. This special funding is often available either for construction of a new water system or specific improvements to an existing system. The principal federal funding sources that may be available to public water systems are from the state revolving fund loans and the Rural Development Administration in the United States.

Private and investor-owned systems are not generally eligible to receive the grants or low-interest loans available to publicly owned systems. As in any other form of private business, the private system owners must create their own financing, usually bond funds, which are taxable. Smaller investor-owned systems must be operated efficiently and must continuously show a good rate of return on investment in order to sustain operations.

Chapter 12

PLANNING FOR THE FUTURE

Local governments throughout North America have a long history of providing water supply and delivery. In part, this commitment is due to the significant investments in infrastructure that must be made, the perception that water is a basic human service, and the use of water supplies as a means to attract, retain, or control growth in a given area. How that is accomplished has both regulatory and policy implications.

The primary purpose of a public water and sewer system, as defined by law, is to provide adequate quantities of reliable service to its customers. Local elected and management officials should realize that water and wastewater utility systems are highly regulated by federal, state/provincial, and local agencies. Local public officials should be cognizant of this fact when making funding and staffing decisions as there are laws that define the responsibilities of a utility system and its operators, which are public health policy driven. To this end, there are a variety of activities that every utility system should pursue, regardless of size. One of these is planning, as upon planning activities lie all other organizational aspects of the operation. Planning is a public-forum activity and provides an indication of what may be expected in the future. It also allows elected officials and staff to outline the need for programs or revisions to the manner in which business is conducted.

Planning should be undertaken on a regular basis by all utility systems to anticipate needs, clarify organizational goals, provide strategic direction for the organization to pursue, and communicate each of these goals to the public. With regard to water systems, it is especially imperative to have continuous planning activities, since many necessary improvements and programs take months or years to implement and/or complete. Without both a short- and a long-term plan to accomplish future needs, the utility will suffer errors in direction, build unnecessary infrastructure, and pursue programs that later are found to provide the wrong information, or level or type of treatment.

Planning can provide a number of long-term benefits: improvements in Insurance Services Office (ISO) ratings for fire insurance rates, renewal of improvements as monies become available, and most importantly, a vision for the system. In creating any plan for a utility system, efforts must be undertaken to understand the operating environment of the utility system. Second, the needs of the water system must be defined, generally from growth projections and analyses of current infrastructure condition from repair records or specific investigations. By funneling this information into the planning process, the result of the effort should be a set of clear goals and definition of objectives. The types of goals and objectives may vary depending on the type of plan developed.

- Strategic plans—action-oriented, management-level plans usually tied to a vision of the system at 5-, 10-, and 20-year windows
- Integrated resource plans—a relatively new concept proposed by the American Water Works Association (AWWA) to incorporate all parts of the community that are affected by water use, including water conservation, multiple sources, watershed protection, wastewater sources, and ecosystem needs over a 20- to 50-year horizon
- Facilities plans—shorter-term plans oriented toward meeting the needs of state revolving fund loan programs
- Master plans—generally 20-year plans that focus on capital construction but may neglect repair and replacement work in favor of expansion

Any utility planning effort should start with a description (and understanding) of the local ecosystem. An understanding of the environment from which water is drawn or into which it is to be discharged is important. Water quality and quantity, whether surface water or groundwater, are profoundly affected by demand. A reduced demand for surface water helps prevent degradation of the quality of the resource in times of low precipitation. Reduction in the pumping of groundwater improves the aquifer's ability to withstand saltwater infiltration, contamination by septic tank leachate, underground storage tank leakage, and leaching of hazardous wastes and other pollutants. Overpumping groundwater leads to contamination of large sections of the aquifer, which are then lost forever as a viable resource. Therefore, groundwater protection must be a part of any planning efforts.

Planning for equipment replacement should be a continuous process. Observation of the equipment through maintenance records, age, and performance are all part of the process for determining what infrastructure needs to be replaced. When a piece of equipment needs to be replaced, investigation should be made of improved models that might be more reliable or efficient. Consideration should also be given to upsizing equipment when it is replaced to allow for future needs. One of the primary advantages of replacing equipment before it completely breaks down is that more time can be taken to evaluate the alternatives before making a selection.

Regulatory issues should also be included in the plan as well as a general description of the service area, including an analysis of the growth patterns and likely new customer bases, including utility acquisitions. Forecasts predicting demands based upon expected growth should be made. Then water-use patterns and projections of variables, such as population and land use, can be used to predict future demands.

An outline of the existing utility system is required, including evaluation of the current asset conditions that led to the needs assessment. Analysis of the current distribution system pressure and water quality will point out problem areas that need main replacement or reinforcements. The distribution system analysis will lend itself to modeling of water quality transport in the future.

Comparisons of water quality between existing and proposed regulations will indicate areas where treatment should be improved. Concepts vital to any plan include means to reduce peak demands (thereby lowering estimates of future capacity needs) and to define when major capital investments are necessary.

It is also necessary to monitor the effectiveness of a program over a long period of time. There are indications that people make changes in short-term water-use behaviors but eventually return to their former habits. This could be disastrous if demand increases drastically over a short period of time and facilities are not on line. This is one reason a public relations program is essential to sustain water conservation efforts. Utilities should experience a general decrease in total operations costs as a result of a properly designed and implemented conservation program.

The result of the plan should be direction for the future of the utility and an identification of operating, capital, and policy needs to meet that direction. Among the more important steps is the capital improvement program. Once the plan is approved, staff and engineers can work on various projects. Included in the process are the following steps:

- Preparation of a study to suggest what should be done and to estimate cost
- Specific decisions by management on work to be done and funding method
- Preparation of plans, specifications, and arrangements made, for funding application for and approval of plans and specifications by the state
- Acceptance of bids and letting of contracts
- Overview of construction to ensure conformity with plans
- Equipment start-up and training of employees

Each of these steps may take months, so it is not unusual for the time between inception and completion of a major project to take several years. Planning, therefore, must be extended as far into the future as possible.

The result of planning should be an outline of activities the utility staff is expected to undertake and an outline for the public of how the goals of the utility will be met, including who pays for the improvements. As a result, the public should expect the following from local elected officials and utility managers:

- The utility will be operated as an enterprise fund so that the costs of the system will be borne by those benefiting from the system without the support of, or providing support to, the general fund operation of local governments.

- Repair and replacement will be ongoing activities to maintain infrastructure in a condition that will ensure reliable service with minimal disruptions.
- Repair and replacement should be an ongoing portion of the budget, rather than, as is too often the case, an afterthought.
- Commercial and industrial enterprises will have significant concerns in this area.

The utility should undertake the following programs as part of its ongoing operations and any facility or financial planning activities:

- Annual fire hydrant testing
- Annual valve exercise program
- Biannual large-user meter repair program
- Ongoing meter change-out for small meters
- Ongoing infiltration and inflow reduction effort in the budget
- Ongoing monitoring and tracking of well and pump performance
- Preventive maintenance of equipment in accordance with manufacturer's suggestions
- Work-order tracking programs to identify areas where maintenance or repair and replacement funding should be allocated
- Water quality testing programs
- Ongoing, appropriate training for staff, such as basic operations, confined space, chemical safety, and license renewals
- Ongoing painting and housekeeping at all facilities to provide a positive public appearance and to maintain public health

Local officials are expected to meet the requirements of all regulations by providing the appropriate staffing, from an educational and experience perspective, and compensating them accordingly to maintain a degree of stability in the system. In addition, local officials are expected to

- Repay all debts
- Budget appropriate monies for operations, repair, replacement, and debt
- Approve rate increases sufficient to meet expenditure needs
- Monitor and report revenues and expenditures on a regular basis

Budgeting appropriate repair and replacement monies is an often overlooked budgetary item. The following are documents that should be on hand in the utility offices:

- Operations and maintenance manuals for all equipment
- Water distribution and sewer collection system maps
- Water and sewer extension policies
- Master plans
- Impact fee ordinances/policies
- Billing procedures
- Capital improvement program

Local officials are expected to plan for emergency situations and require managers to provide ongoing reporting of the utility system operations to maintain public confidence. This should include water quality reports, ongoing maintenance, and treatment plant operations reports. A public relations effort should be in place to communicate the complexities of the utility operations to the public. This is especially important in times of emergencies or drought.

FINAL THOUGHTS

The intention of this text was to provide information on the operation and maintenance of public water supply systems. The water industry has become more regulated with emphasis on customer safety and water quality as a result of the water quality regulations placed on water and sewer systems. Public health and safety are inherent in everything utility managers do.

The technology required for safe operation of water systems has increased substantially. As a result, many water system issues require the use of professionals and consultants. There are two outcomes from improvements in technology: safer water for the customer and higher costs of operation of the water system. However, in most communities, *water is still less expensive than cable television* and water is essential for our survival. Water is a bargain in North America; its low cost makes it accessible by everyone.

The future will bring additional water quality and supply issues to the fore. Water sources continue to have threats from industrial pollution, septic systems, agricultural runoff, and stormwater. While point sources like wastewater discharges and industrial discharges are easy to regulate and monitor, contaminants from nonpoint sources like agricultural operations may yet contaminate water supplies, if only by accident. Nonpoint sources are nearly impossible to monitor. Agricultural runoff contains animal waste, fertilizers, herbicides, pesticides, and endocrine disruptors in the form of hormones and antibiotics. Endocrine disruptors are the next wave requiring regulation due to their suspected links to cancer and gender confusion in aquatic species. Treatment of these constituents is relatively unstudied.

Finally, a utility system will continue to age. Reinvestment in infrastructure is lacking in most utility systems. As a result, many sources indicate that the nation's water systems can expect massive expenditures for infrastructure replacement in the coming 50 years as pipes wear out. Investment in upgrades to treatment systems, new treatment alternatives, and mechanical upgrades will be ongoing capital expenditures, as they have been in the past. Making these improvements while ensuring access to adequate water for all residents will be a public policy challenge for the industry.

ADDITIONAL SOURCES OF INFORMATION

Appendix A of this book lists organizations that have additional information available on various aspects of public water system operation and management. Public officials are encouraged to call or write to these organizations to obtain current publication lists and copies of publications relating to their particular water system and current interests.

Appendix B lists the contact information of the US Environmental Protection Agency Drinking Water Program regional offices and the states included within each region.

Appendix C lists the addresses of agencies in each state responsible for public water supply overview. Public officials should contact their state agency for a copy of all regulations and other available information on water system operations. Information on water system operations and management can also be obtained from officials and water supply professionals in larger water systems, consulting engineers, state public water supply program personnel, and local water operator groups.

Many publications detail public water system design and operations. Many are available from AWWA, as this handbook is. Appendix D lists several published resources, but the list is not complete due to the large number available and the frequency of revision of same due to constantly changing regulations and technological improvements. Publications that are more than a few years old may no longer be accurate.

REFERENCES

AWWA. 2003. *AWWA Policy Statement on Operator Certification.* Denver, Colo.: American Water Works Association. www.awwa.org/policystatements.

AWWA. 2004a. *AWWA Policy Statement on Metering and Accountability.* Denver, Colo.: American Water Works Association. www.awwa.org/policystatements.

AWWA. 2004b. *AWWA Policy Statement on Safety.* Denver, Colo.: American Water Works Association. www.awwa.org/policystatements.

AWWA. 2005a. *AWWA Policy Statement on Employee Compensation.* Denver, Colo.: American Water Works Association. www.awwa.org/policystatements.

AWWA. 2005b. *AWWA Policy Statement on Financing, Accounting, and Rates.* Denver, Colo.: American Water Works Association. www.awwa.org/policystatements.

AWWA. 2005c. *AWWA Policy Statement on Cross Connections.* Denver, Colo.: American Water Works Association. www.awwa.org/policystatements.

AWWA. 2007. *AWWA Policy Statement on Public Involvement and Customer Communication.* Denver, Colo.: American Water Works Association. www.awwa.org/policystatements.

AWWA. 2008a. *AWWA Policy Statement on Discontinuance of Water Service for Nonpayment.* Denver, Colo.: American Water Works Association. www.awwa.org/policystatements.

AWWA. 2008b. *AWWA Policy Statement on Employee Training and Career Development.* Denver, Colo.: American Water Works Association. www.awwa.org/policystatements.

Blackstone, E. A., and S. Hakim. 1997. Private Ayes: A Tale of Four Cities. *American City and County* 112 no. 2 (February):PS4–PS12.

Bloetscher, F., A. Muniz, and J. Largey. 2007. *Siting, Drilling, and Construction of Water Supply Wells.* Denver, Colo.: American Water Works Association.

Bloetscher, F., A. Muniz, and G. M. Witt. 2005. *Groundwater Injection: Modeling, Risks, and Regulations.* New York: McGraw-Hill.

Bloetscher, Frederick, and Richard W. Saltrick. 1998. Integrated Resource Planning Helps the City of Hollywood Insure Its Long-Term Water Supplies. In *Florida Section American Water Works Association Annual Conference Proceedings,* Orlando, Fla.

Broward County Board of County Commissioners. 2001. *Broward County Wellfield Map* and *What are Zones of Influence?* http://www.broward.org/pprd/wf_map.pdf and http://www.broward.org/pprd/wf5.htm.

City of Dunedin v. Contractors and Builders Association of Pinellas County, 312 So. 2d 763. (Fla. 2d DCA 1975).

City of Miami v. Florida Public Service Commission, 208 So. 2d 249 (Fla. 1968).

Eaton, A., L. Clesceri, E. Rice, and A. E. Greenburg. 2005. *Standard Methods for the Examination of Water and Wastewater,* 21st ed. Denver, Colo.: American Public Health Association, Water Environment Federation, and American Water Works Association.

Florida Department of Environmental Protection. 2006. Reuse of Reclaimed Water and Land Application. Chapter 62-610, *Florida Administrative Code.* www.dep.state.fl.us/legal/Rules/wastewater/62-610.pdf.

Hollywood, Inc. v. Broward County. 1983. 431 So. 2d 606. District Court of Appeals of Florida, Fourth District, March 23.

Miami–Dade County. 2006. *Typical Fire Hydrant Installation.* www.miamidade.gov/wasd/library/Donation/part-4/PDF/WS_4_50_2OF-03-14-07.pdf.

Miami–Dade County. 2006. *Typical 1" Service Connection.* www.miamidade.gov/wasd/library/Donation/part-4/PDF/WS_4.11_2.pdf.

O'Connor, D. 2002. *Walkerton Commission of Inquiry.* www.attorneygeneral.jus.gov.on.ca/english/about/pubs/walkerton.

Appendix A

ORGANIZATIONS WITH ADDITIONAL INFORMATION AND ASSISTANCE

USEPA

The US Environmental Protection Agency (USEPA) is responsible for protection of drinking water supplies under the Safe Drinking Water Act. The agency establishes national drinking water standards and monitors state enforcement of drinking water standards, system management, and operations. The USEPA's principal office is located in Washington, D.C. The agency maintains a Safe Drinking Water Hotline to provide information on drinking water regulations, policies, and documents. The hotline hours are 8:30 a.m. to 5:00 p.m. Eastern Standard Time, Monday through Friday, excluding holidays. The hotline number is (800) 426-4791.

There are also 10 regional offices of the USEPA in the United States. These offices may be contacted for information on drinking water regulations and policies regarding water systems located within their region. The addresses and phone numbers of the regional offices are listed in appendix B.

State Drinking Water Agencies

State drinking water agencies have been designated by the governor of each state to accept primary enforcement responsibility for the operation of the program within their state. Each agency also has several field offices that can be contacted for specific information on state requirements for the operation of public water systems. The principal offices of each state agency are listed in appendix C.

American Water Works Association

The American Water Works Association (AWWA) is a scientific and educational association that conducts research and provides technical publications, information, training, and technical assistance to the public water supply industry. The central office is in Denver, Colo. Each state is also represented by an AWWA section that is active in holding meetings and presenting training classes. For further information, contact

American Water Works Association
6666 W. Quincy Avenue
Denver, CO 80235
(303) 794-7711

National Rural Water Association

The National Rural Water Association (NRWA) provides technical publications, training, and technical assistance to small water systems and rural water districts. Many states also have a state organization and staff who provide technical assistance and training. For further information, contact

National Rural Water Association
P.O. Box 1428
Duncan, OK 73534
(405) 252-0629

Rural Community Assistance Partnership

The Rural Community Assistance Partnership (RCAP) consists of six regional agencies formed to develop the capacity of rural community officials to solve local water problems. The program provides on-site technical assistance, training, and publications to rural communities. The addresses of RCAP agency offices are listed in appendix C.

National Ground Water Association

The National Ground Water Association is a not-for-profit professional society and trade association representing the groundwater industry. The association provides expositions, education, and research on wells and groundwater and has many publications available on groundwater subjects. For information, contact

National Ground Water Association
601 Dempsey Road
Westerville, OH 43081
(800) 551-7379
(614) 898-7791

New England Water Works Association

The New England Water Works Association (NEWWA) is a membership organization representing consultants, water supply operations and management professionals, and technical experts. NEWWA sponsors workshops, offers publications, and provides its members with an opportunity to exchange ideas and information on water works operations and management.

New England Water Works Association
125 Hopping Brook Road
Holliston, MA 01746
(508) 893-7979

Rural Development Administration

The Rural Development Administration (RDA) provides grants and loans for rural water systems and communities with populations less than 25,000. For additional information, contact

Rural Development Administration
14th and Independence Avenue SW
Washington, DC 20250
(202) 720-9619

National Drinking Water Clearinghouse

The National Drinking Water Clearinghouse (NDWC) was established in 1991 at West Virginia University to develop and maintain services and information related to small community drinking water systems. Intended for communities of less than 10,000 people and those who work with them, the NDWC provides publications, databases, referrals, and educational products.

National Drinking Water Clearinghouse
West Virginia University
P.O. Box 6064
Morgantown, WV 26506-6064
(800) 624-8301

USEPA HEADQUARTERS

Standard Mailing Address

US Environmental Protection Agency
Ariel Rios Building
1200 Pennsylvania Avenue NW
Washington, DC 20460
(202) 272-0167

Overnight Package Delivery Mailing Address

US Environmental Protection Agency
USEPA East
1201 Constitution Avenue NW
Washington, DC 20004

This page intentionally blank.

Appendix B

US ENVIRONMENTAL PROTECTION AGENCY REGIONAL OFFICES

Region 1 (CT, MA, ME, NH, RI, VT)
Environmental Protection Agency
1 Congress St. Suite 1100
Boston, MA 02114-2023
www.epa.gov/region01/
Phone: (617) 918-1111
Fax: (617) 565-3660
Toll free within Region 1:
(888) 372-7341

Region 2 (NJ, NY, PR, VI)
Environmental Protection Agency
290 Broadway
New York, NY 10007-1866
www.epa.gov/region02/
Phone: (212) 637-3000
Fax: (212) 637-3526
E-mail: R2_Web_Inquiry@epamail.epa.gov.

Region 3 (DC, DE, MD, PA, VA, WV)
Environmental Protection Agency
1650 Arch Street
Philadelphia, PA 19103-2029
www.epa.gov/region03/
Phone: (215) 814-5000
Fax: (215) 814-5103
Toll free: (800) 438-2474
E-mail: r3public@epa.gov

Region 4 (AL, FL, GA, KY, MS, NC, SC, TN)
Environmental Protection Agency
Atlanta Federal Center
61 Forsyth Street, SW
Atlanta, GA 30303-3104
www.epa.gov/region04/
Phone: (404) 562-9900
Fax: (404) 562-8174
Toll free: (800) 241-1754

Region 5 (IL, IN, MI, MN, OH, WI)
Environmental Protection Agency
77 West Jackson Boulevard
Chicago, IL 60604-3507
www.epa.gov/region5/
Phone: (312) 353-2000
Fax: (312) 353-4135
Toll free within Region 5:
(800) 621-8431

Region 6 (AR, LA, NM, OK, TX)
Environmental Protection Agency
Fountain Place 12th Floor, Suite 1200
1445 Ross Avenue
Dallas, TX 75202-2733
www.epa.gov/region06/
Phone: (214) 665-2200
Fax: (214) 665-7113
Toll free within Region 6:
(800) 887-6063

Region 7 (IA, KS, MO, NE)

Environmental Protection Agency
901 North 5th Street
Kansas City, KS 66101
www.epa.gov/region07/
Phone: (913) 551-7003
Toll free: (800) 223-0425
E-mail: r7actionline@epa.gov.

Region 8 (CO, MT, ND, SD, UT, WY)

Environmental Protection Agency
999 18th Street Suite 500
Denver, CO 80202-2466
www.epa.gov/region08/
Phone: (303) 312-6312
Fax: (303) 312-6339
Toll free: (800) 227-8917
E-mail: r8eisc@epa.gov

Region 9 (AZ, CA, HI, NV, GU, MP)

Environmental Protection Agency
75 Hawthorne Street
San Francisco, CA 94105
www.epa.gov/region09/
Phone: (415) 947-8000
Toll free within Region 9:
(866) EPA-WEST
Fax: (415) 947-3553
E-mail: r9.info@epa.gov

Region 10 (AK, ID, OR, WA)

Environmental Protection Agency
1200 Sixth Avenue
Seattle, WA 98101
www.epa.gov/region10/
Phone: (206) 553-1200
Fax: (206) 553-0149
Toll free: (800) 424-4372

Appendix C

STATE DRINKING WATER AGENCIES AND RURAL COMMUNITY ASSISTANCE PROGRAM AGENCIES

REGION 1

Connecticut Dept. of Public Health
MS#51 WAT
P.O. Box 340308
Hartford, CT 06134
Phone: (860) 509-7343
Fax: (860) 509-7359

Connecticut Dept. of Environmental Protection
Water Management Bureau
79 Elm Street
Hartford, CT 06106-5127
Phone: (860) 424-3718

Massachusetts Dept. of Environmental Protection
Drinking Water Program
One Winter Street
Boston, MA 02108
Phone: (617) 556-1157

Maine Drinking Water Program
Bureau of Health, Div. of Health Engineering
10 State House Station
Augusta, ME 04333-0010
Phone: (207) 287-6196
Fax: (207) 287-4172

New Hampshire Dept. of Environmental Services
Water Supply Engineering Bureau
6 Hazen Drive, P.O. Box 95
Concord, NH 03302
Phone: (603) 271-1168
Fax: (603) 271-2181

Rhode Island Dept. of Health
Office of Drinking Water Quality
3 Capitol Hill
Providence, RI 02908-5097
Phone: (401) 222-7769
Fax: (401) 222-6953

Rhode Island Dept. of Environment
235 Promenade Street
Providence, RI 02908
Phone: (401) 222-2234 X7603

Vermont Dept. of Environmental Conservation
103 S. Main Street
Waterbury, VT 06571-0403
Phone: (802) 241-3418
Fax: (802) 241-3284

REGION 2

New Jersey Dept. Environmental Protection
Bureau of Safe Drinking Water,
CN 426
401 E. State Street
Trenton, NJ 08625-0426
Phone: (609) 292-5550
Fax: (609) 292-1654

New York State Dept. of Health
Bureau of Public Water Supply Protection
Flanigan Square
547 River Street, Room 400, 4th floor
Troy, NY 12180-2216
Phone: (518) 402-7650
Fax: (518) 402-7599

Puerto Rico Dept. of Health
Public Water Supervision Program
P.O. Box 70184
Edificio A. Centro Medico
San Juan, PR 00909
Phone: (787) 754-6370
Fax: (787) 754-6010

Planning and Natural Resources
Government of Virgin Islands
Nitky Center, Suite 231
St. Thomas, Virgin Islands 00802
(809) 774-3320

REGION 3

Delaware Water Supply Section
Division of Water Resources
Delaware Dept. of Natural Resources and Environmental Control
P.O. Box 1401
Dover, DE 19903
Phone: (302) 739-4793
Fax: (302) 739-2296

Office of Sanitary Engineering
Delaware Division of Public Health
Cooper Building
P.O. Box 637
Dover, DE 19903
Phone: (302) 739-5410

Water Quality Division
DC Department of Health
51 N Street, NE, 5th Floor
Washington, DC 20002
Phone: (202) 535-1876
Fax: (202) 535-1632

Water Supply Program
Maryland Department of the Environment
Water Management Administration
Montgomery Park Business Center
1800 Washington Boulevard
Baltimore, MD 21230
Phone: (410) 537-3714
Fax: (410) 537-3157
Toll free: (800) 633-6101

Department of Environmental Protection
Bureau of Water Supply and
Wastewater Management
P.O. Box 8467
Harrisburg, PA 17105-8467
Phone: (717) 783-3795

Virginia Department of Health
P.O. Box 2448
1500 East Main Street
Richmond, VA 23218-2448

Virginia Department of
Environmental Quality
629 East Main Street
P.O. Box 10009
Richmond, VA 23240-0009
Phone: (804) 698-4000

West Virginia Dept. of Health
Environmental Engineering Division
815 Quarrier Street, Suite 418
Charleston, WV 25301
Phone: (304) 558-2981
Fax: (304) 558-0691

West Virginia Dept. of Environmental
Protection
414 Summers Street
Charleston, WV 25301
Phone: (304) 558-2107
Fax: (304) 558-5905

REGION 4

Department of Environmental
Management
1400 Coliseum Drive
P.O. Box 301463
Montgomery, AL 36130-1463
Phone: (334) 271-7832
Fax: (334) 271-7950

Bureau of Water Resources
Protection
Dept. of Environmental Protection
Twin Towers Office Building
2600 Blair Stone Road
Tallahassee, FL 32399-2400
Phone: (850) 245-8645

Georgia Environmental Protection
Division
Suite 1362 E. Floyd Towers
205 Butler Street, SE
Atlanta, GA 30334
Phone: (404) 656-0719
Fax: (404) 651-9590

Kentucky Division of Water
Natural Resources and Environmental
Protection Cabinet
14 Reilly Road
Frankfort, KY 40601
Phone: (502) 564-3410

State of Mississippi Ground Water
Planning Branch
P. O. Box 10385
Jackson, MS 39289-0385
Phone: (601) 961-5395

Public Water Supply System
North Carolina Dept. of Environment
and Natural Resources
1634 Mail Service Center
Raleigh, NC 27699-1634
Phone: (919) 715-3224
Fax: (919) 715-4374

Ground Water Management Section
South Carolina Dept. of Health and
Environmental Control
2600 Bull Street
Columbia, SC 29201-1708
Phone: (803) 898-3798
Fax: (803) 898-4190

Ground Water Management Section
Division of Water Supply
Dept. of Environment and
Conservation
401 Church Street
Nashville, TN 37243-1549
Phone: (615) 532-0170

REGION 5

Division of Public Water Supplies
Illinois Environmental Protection Agency
P.O. Box 19276
Springfield, IL 62794-9276
Phone: (217) 785-4787
Fax: (217) 782-0075

Indiana Dept. of Environmental Management
P.O. Box 6015
Indianapolis, IN 46206-6015
Phone: (317) 308-3388
Fax: (317) 308-3339

Drinking Water and Radiological Division
Michigan Dept. of Environmental Quality
P.O. Box 30630
Lansing, MI 48909-8130
Phone: (517) 241-1359
Fax: (517) 241-1328

Minnesota Drinking Water Protection Section
Minnesota Dept. of Health
P.O. Box 64975
St. Paul, MN 55164-0975
Phone: (651) 201-4610

Division of Drinking and Ground Waters
Ohio Environmental Protection Agency
P.O. Box 1049
Columbus, OH 43216-1049
Phone: (614) 644-2752
Fax: (614) 644-2909

Wisconsin Dept. of Natural Resources
Bureau of Drinking Water and Groundwater
P.O. Box 7921
Madison, WI 53707-7921
Phone: (608) 266-5234
Fax: (608) 267-7650

REGION 6

Arkansas Dept. of Health
Division of Engineering
4815 W. Markham Street, Mail Slot 37
Little Rock, AR 72205-3867
Phone: (501) 661-2623
Fax: (501) 661-2032

Louisiana Dept. of Environmental Quality
Environmental Evaluation Division
Aquifer Evaluation and Protection Section
P.O. Box 4314
Baton Rouge, LA 70821-4314
Phone: (225) 765-0578

Drinking Water Bureau
New Mexico Environment Dept.
525 Camino de los Marquez, Suite 4
Santa Fe, NM 87501
Phone: (505) 827-1400

Water Quality Protection Division
Oklahoma Dept. of Environmental Quality
P.O. Box 1677
Oklahoma City, OK 73101
Phone: (405) 702-8120

Public Drinking Water Section (MC-155)
Texas Commission on Environmental Quality
P.O. Box 13087
Austin, TX 78711-3087
Phone: (512) 239-6020
Fax: (512) 239-6050

REGION 7

Iowa Department of Natural Resources
401 SW 7th Street, Suite M
Des Moines, IA 50309-4611
Phone: (515) 725-0275
Fax: (515) 725-0348

Kansas Department of Health and Environment
Watershed Management Section
Bureau of Water
1000 SW Jackson Street, Suite 420
Topeka, KS 66612-1367
Phone: (785) 296-5535

Public Drinking Water Program
Missouri Department of Natural Resources
Division of Environmental Quality
P.O. Box 176
Jefferson City, MO 65102-0176
Phone: (573) 526-5448

Nebraska Department of Environmental Quality
Planning Unit, Water Quality Division
Suite 400, The Atrium
1200 N Street
Lincoln, NE 68509-8922
Phone: (402) 471-4270

REGION 8

Colorado Dept. of Public Health and Environment
WQCD-OA-B2
4300 Cherry Creek Drive South
Denver, CO 80246-1530
Phone: (303) 692-3579
Fax: (303) 782-0390

Montana Dept. of Environmental Quality
Metcalf Building, Box 200901
Helena, MT 59620-0901
Phone: (406) 444-4806
Fax: (406) 444-1374

Division of Water Quality
1200 Missouri Avenue
Bismarck, ND 58504
Phone: (701) 328-5233
Fax: (701) 328-5200

North Dakota Department of Health
P.O. Box 5520
Bismarck, ND 58502-5520
Phone: (701) 328-5241
Fax: (701) 328-5200

South Dakota Dept. of Environment
Joe Foss Building
523 East Capitol
Pierre, SD 57501-3181
Phone: (605) 773-3296
Fax: (605) 773-6035

Utah Dept. of Environmental Quality
Division of Drinking Water
P.O. Box 144830
150 North 1950 West
Salt Lake City, UT 84114-4830
Phone: (801) 536-4195
Fax: (801) 536-4211

Wyoming Dept. of Environmental Quality
Water Quality Division
122 West 25th Street
Herschler Building, 4W
Cheyenne, WY 82002
Phone: (307) 777-7343
Fax: (303) 777-5973

REGION 9

Drinking Water Section
Water Quality Division
Arizona Dept. of Environmental Quality
1110 W. Washington Street MC 5415B-2
Phoenix, AZ 85007
Phone: (602) 771-4641
Fax: (602) 207-4634

Department of Health Services
Drinking Water Technical Programs Branch
50 D Street, Suite 200
Santa Rosa, CA 95404
Phone: (707) 576-2295
Fax: (707) 576-2722

Rural Community Assistance Corporation
3120 Freeboard Drive, Suite 201, 2nd Floor
West Sacramento, CA 95691
Phone: (916) 447-2854
Fax: (916) 447-2878

Water Resource Management Program
Guam Environmental Protection Agency
P.O. Box 22439
Guam Main Facility
Barrigada, Guam 96921
Phone: (671) 475-1641
Fax: (671) 477-9402

Safe Drinking Water Branch
Hawaii Dept. of Health
919 Ala Moana Blvd., Rm. 308
Honolulu, HI 96814
Phone: (808) 586-4258
Fax: (808) 586-4370

Nevada Bureau of Health Protection Services
1179 Fairview Drive, Ste. 201
Carson City, NV 89701-5405
Phone: (775) 687-4754
Fax: (775) 687-5699

Nevada Division of Environmental Protection
333 West Nye Lane, Suite 138
Carson City, NV 89706-0851
Phone: (775) 687-9426
Fax: (775) 687-4684

Northern Mariana Islands
CNMI Division of Environmental Quality
Drinking Water Program
P.O. Box 1304-CK
Saipan, MP 96950
Phone: (670) 234-1012
Fax: (670) 234-1003

REGION 10

Drinking Water Protection Program
Division of Environmental Health
555 Cordova Street
Anchorage, AK 99501
Phone: (907) 269-7521
Fax: (907) 269-3990

Drinking Water Program
Alaska Dept. of Environmental Conservation
555 Cordova Street
Anchorage, AK 99501
Phone: (907) 269-7685

Idaho Dept. of Environmental Quality
1410 North Hilton
Boise, ID 83706
Phone: (208) 373-0274
Fax: (208) 373-0576

Oregon Dept. of Environmental Quality
811 SW 6th Avenue
Portland, OR 97204-1390
Phone: (503) 229-5413
Fax: (503) 229-5408

Oregon Groundwater Coordinator
Drinking Water Protection Program
442 A Street
Springfield, OR 97477
Phone: (541) 726-2587
Fax: (541) 726-2596

Division of Drinking Water, Dept. of Health
P.O. Box 47849
Olympia, WA 98504-7849
Phone: (360) 236-3149
Fax: (360) 236-2254

OTHER AGENCIES

Great Lakes Rural Network
P.O. Box 590
Fremont, OH 43420
Phone: (419) 334-5124
Toll free: (800) 775-9767

Midwest Assistance Program, Inc.
P.O. Box 81
212 Lady Slipper Ave. NE
New Prague, MN 56071
Phone: (952) 758-4334
Fax: (952) 758-4336
E-mail: plmap@bevcomm.net

Rural Community Assistance Partnership
RCAP, Inc.
1522 K Street NW, Suite 400
Washington, DC 20005
Phone: (202) 408-1273
Toll free: (888) 321-7227
Fax: (202) 408-8165

Rural Housing Improvement, Inc.
218 Central Street, Box 429
Winchendon, MA 01475-0429
Phone: (617) 297-1376

Rural Development: United States Department of Agriculture (USDA)
NOTE: All applications for loans and grants are handled at the local level. For help with an application, contact your Rural Development State Office or Rural Development staff at your nearest USDA Service Center.

Office of the Under Secretary
USDA Rural Development,
Room 206-W
Mail Stop 0107
1400 Independence Avenue SW
Washington, DC 20250-0107
Phone: (202) 720-4581
TTY: (800) 877-8339 (Federal Information Relay Service)
Fax: (202) 720-2080 www.rurdev.usda.gov/scrty/index.html

Legislative and Public Affairs Staff
USDA Rural Development,
Room 4801-S
Mail Stop 0705
1400 Independence Avenue SW
Washington, DC 20250-0705
Phone: (202) 720-4323
TTY: (800) 877-8339
(Federal Information Relay Service)
Fax: (202) 690-4083

Rural Housing Service (RHS)
National Office
USDA/RHS, Room 5014-S
Mail Stop 0701
1400 Independence Avenue SW
Washington, DC 20250-0701
Phone: (202) 690-1533
TTY: (800) 877-8339 (Federal Information Relay Service)
Fax: (202) 690-0500 www.rurdev.usda.gov/rhs/index.html

Rural Housing Service
Centralized Servicing Center
1520 Market Street
St. Louis, MO 63103
Phone: (800) 414-1226 (Toll Free)
TTY: (800) 438-1832 (Toll Free)
Fax: (314) 206-2805 www.rurdev.usda.gov/rhs/sfh/bor_sfh.htm

Rural Business-Cooperative Service (RBS)
National Office
USDA/RBS, Room 5045-S
Mail Stop 3201
1400 Independence Avenue SW
Washington, DC 20250-3201
Phone: (202) 690-4730
TTY: (800) 877-8339 (Federal Information Relay Service)
Fax: (202) 690-4737 www.rurdev.usda.gov/rbs/

Rural Utilities Service (RUS)
National Office
USDA/RUS, Room 4051-S
Mail Stop 1510
1400 Independence Avenue SW
Washington, DC 20250-1510
Phone: (202) 720-9540
TTY: (800) 877-8339 (Federal Information Relay Service)
Fax: (202) 720-1725
www.usda.gov/rus/

Office of Community Development (OCD)
National Office
USDA/OCD, Room 266
Mail Stop 3203
300 7th Street SW
Washington, DC 20250-3203
Phone: (202) 619-7980
TTY: (800) 877-8339 (Federal Information Relay Service)
Fax: (202) 401-7420

http://ocdweb.sc.egov.usda.gov/
USDA Washington Telephone Directory (Includes Rural Development)
http://dc-directory.hqnet.usda.gov/DLSNew/phone.aspx

State Office Mailing Addresses
www.rurdev.usda.gov/scrty/sdirs.html

Information is current as of October 2008.

Appendix D

ADDITIONAL PUBLISHED RESOURCES

ANSI/AWWA. Standard C104/A21.4: Cement–Mortar Lining for Ductile-Iron Pipe and Fittings. Denver, Colo.: American Water Works Association.

ANSI/AWWA. Standard C400: Asbestos–Cement Pressure Pipe, 4 In. Through 16 In. (100 mm Through 400 mm), for Water Distribution Systems. Denver, Colo.: American Water Works Association. [withdrawn]

ANSI/AWWA. Standard C500: Metal-Seated Gate Valves for Water Supply Service. Denver, Colo.: American Water Works Association.

ANSI/AWWA. Standard C502: Dry-Barrel Fire Hydrants. Denver, Colo.: American Water Works Association.

ANSI/AWWA. Standard C503: Wet-Barrel Fire Hydrants. Denver, Colo.: American Water Works Association.

ANSI/AWWA. Standard C508: Swing-Check Valves for Waterworks Service, 2 In. (50 mm) Through 24 In. (600 mm) NPS. Denver, Colo.: American Water Works Association.

ANSI/AWWA. Standard C509: Resilient-Seated Gate Valves for Water Supply Service. Denver, Colo.: American Water Works Association.

ANSI/AWWA. Standard C510: Double Check Valve Backflow-Prevention Assembly. Denver, Colo.: American Water Works Association.

ANSI/AWWA. Standard C511: Reduced-Pressure Principle Backflow-Prevention Assembly. Denver, Colo.: American Water Works Association.

ANSI/AWWA. Standard C512: Air Release, Air/Vacuum, and Combination Air Valves for Waterworks Service. Denver, Colo.: American Water Works Association.

ANSI/AWWA. Standard C540: Power-Actuating Devices for Valves and Slide Gates. Denver, Colo.: American Water Works Association.

ANSI/AWWA. Standard C700: Cold-Water Meters—Displacement Type, Bronze Main Case. Denver, Colo.: American Water Works Association.

ANSI/AWWA. Standard C702: Cold-Water Meters—Compound Type. Denver, Colo.: American Water Works Association.

ANSI/AWWA. Standard C704: Propeller-Type Meters for Waterworks Applications. Denver, Colo.: American Water Works Association.

ANSI/AWWA. Standard C710: Cold-Water Meters—Displacement Type, Plastic Main Case. Denver, Colo.: American Water Works Association.

ANSI/AWWA. Standard C800: Underground Service Line Valves and Fittings. Denver, Colo.: American Water Works Association.

ANSI/AWWA. Standard C900: Polyvinyl Chloride (PVC) Pressure Pipe, 4 In. Through 12 In. (100 mm Through 300 mm), for Water Distribution. Denver, Colo.: American Water Works Association.

ANSI/AWWA. Standard C901: Polyethylene (PE) Pressure Pipe and Tubing, ½ In. (13 mm) Through 3 In. (76 mm), for Water Service. Denver, Colo.: American Water Works Association.

ANSI/AWWA. Standard C903: Polyethylene–Aluminum–Polyethylene and Cross-linked Polyethylene–Aluminum–Cross-linked Polyethylene Composite Pressure Pipes, ½ In. (12 mm) through 2 In. (50 mm), for Water Service. Denver, Colo.: American Water Works Association.

ANSI/AWWA. Standard C906: Polyethylene (PE) Pressure Pipe and Fittings, 4 In. (100 mm) Through 63 In. (1,600 mm), for Water Distribution and Transmission. Denver, Colo.: American Water Works Association.

ANSI/AWWA. Standard C907: Injection-Molded Polyvinyl Chloride (PVC) Pressure Fittings, 4 In. Through 12 In. (100 mm through 300 mm), for Water Distribution. Denver, Colo.: American Water Works Association.

AWWA. 1991. *Guidance Manual for Compliance with the Filtration and Disinfection Requirements for Public Water Systems Using Surface Water.* Denver, Colo.: American Water Works Association.

AWWA. Manual M1: *Principles of Water Rates, Fees, and Charges.* Denver, Colo.: American Water Works Association.

AWWA. Manual M3: *Safety Practices for Water Utilities.* Denver, Colo.: American Water Works Association.

AWWA. Manual M5: *Water Utility Management.* Denver, Colo.: American Water Works Association.

AWWA. Manual M6: *Water Meters—Selection, Installation, Testing, and Maintenance.* Denver, Colo.: American Water Works Association.

AWWA. Manual M14: *Recommended Practice for Backflow Prevention and Cross-Connection Control.* Denver, Colo.: American Water Works Association.

AWWA. Manual M17: *Installation, Field Testing, and Maintenance of Fire Hydrants.* Denver, Colo.: American Water Works Association.

AWWA. Manual M19: *Emergency Planning for Water Utilities.* Denver, Colo.: American Water Works Association.

AWWA. Manual M21: *Groundwater.* Denver, Colo.: American Water Works Association.

AWWA. Manual M22: *Sizing Water Service Lines and Meters.* Denver, Colo.: American Water Works Association.

AWWA. Manual M28: *Rehabilitation of Water Mains.* Denver, Colo.: American Water Works Association.

AWWA. Manual M29: *Water Utility Capital Financing.* Denver, Colo.: American Water Works Association.

AWWA. Manual M31: *Distribution System Requirements for Fire Protection.* Denver, Colo.: American Water Works Association.

AWWA. Manual M37: *Operational Control of Coagulation and Filtration Processes.* Denver, Colo.: American Water Works Association.

AWWA. Manual M44: *Distribution Valves—Selection, Testing, Installation, and Maintenance.* Denver, Colo.: American Water Works Association.

AWWA. Manual M46: *Reverse Osmosis and Nanofiltration.* Denver, Colo.: American Water Works Association.

AWWA. Manual M48: *Waterborne Pathogens.* Denver, Colo.: American Water Works Association.

AWWA. Manual M50: *Water Resources Planning.* Denver, Colo.: American Water Works Association.

AWWA. Standard C905: Polyvinyl Chloride (PVC) Pressure Pipe and Fabricated Fittings, 14 In. Through 48 In. (350 mm Through 1,200 mm), for Water Transmission and Distribution. Denver, Colo.: American Water Works Association.

Bloetscher, Frederick. "Deferred Maintenance Obligations Due to Aging Utility Infrastructure." In *Annual Conference Proceedings, Water Environment Federation Technical Exhibition and Conference,* Orlando, Fla., 1999.

Bloetscher, Frederick. "Issues of Concern in the Evaluation of Any Service Delivery Mechanism." In *Annual Conference Proceedings, Florida Section American Water Works Association,* Orlando, Fla., 1999.

Bloetscher, Frederick. "The Long-Term Benefits of Establishing Reserve Funds." In *Florida Water Resource Conference Proceedings,* Kissimmee, Fla., 2004.

Bloetscher, Frederick, and Robert E. Fergen. "Diversified Water Supply Assures Meeting Future Needs of Coastal Community." In *9th South Carolina Environmental Conference Proceedings,* Myrtle Beach, S.C., 1999.

Bloetscher, Frederick, and Robert E. Fergen. "Are Pharmaceutically Active Substances (PASs) the Next Major Water Pollution Issue?" In *Florida Section American Water Works Association Annual Conference Proceedings,* Orlando, Fla., 2001.

Bloetscher, Frederick, and Robert E. Fergen. "The Next Generation of Regulations: Pharmaceuticals as Endocrine Disruptors." In *Annual Conference Proceedings, Environmental Water Resources Institute, American Society of Civil Engineers,* Reston, Va., 2002.

Bloetscher, Frederick, W. L. Jarocki, and P. Varney. "Reserve Funds vs. Borrowing: The Effects on Customer Rates." In *EWRI Conference Proceedings,* Salt Lake City, Utah, 2004.

Bloetscher, Frederick, William Lynch, and Thomas F. Boyd. "A Comparison of Public and Private Utility Service Mechanisms." *Florida Water Resource Journal* 53, no. 4 (April 2001): 44–46.

Bloetscher, Frederick, and Roberto S. Ortiz. "The City of Hollywood's Experience with Multi-Source Membrane Treatment Processes." *Florida Water Resource Journal* 51 no. 1 (January 1999): 24–25.

Bloetscher, Frederick, and Gerhardt M. Witt. "Effects of Silt, Salt, and Other Fouling Mechanisms Affecting Groundwater Treatability." In *American Water Works Association Annual Conference and Exposition Proceedings,* Denver, Colo., 2000.

Bloetscher, Frederick, Gerhardt M. Witt, and Robert E. Fergen. "Biofouling in Raw Water Supply Wells and Its Impact." *Water Engineering and Management* 148, no. 10 (2001): 12–16.

Great Lakes and Upper Mississippi River Board of State Public Health and Environmental Managers (GLUMB). *Recommended Standards for Water Works.* Albany, N.Y.: GLUMB, 1992.

Hydraulic Institute. *Engineering Data Book.* Cleveland, Ohio: Hydraulic Institute, 1990.

US Environmental Protection Agency. *Guidance Manual for Conducting Sanitary Surveys of Public Water Systems; Surface Water and Ground Water Under the Direct Influence (GWUDI).* EPA-815-R-99-016. Washington, D.C.: US Environmental Protection Agency, Office of Water, 1999.

US Environmental Protection Agency. *Cross-Connection Manual.* EPA-816-R-03-002. Washington, D.C.: US Environmental Protection Agency, 2003.

INDEX

Note: *f.* indicates figure;
t. indicates table.

This page intentionally blank.

ABOUT THE AUTHOR

Frederick Bloetscher, PhD, PE, is an assistant professor in the Civil Engineering Department at Florida Atlantic University (FAU) in Boca Raton. FAU is part of the state university system, with more than 28,000 students. At FAU, he teaches classes in water and wastewater treatment, fate and transport of pollutants, water resource engineering, hydrology, and modeling. He also co-teaches the department's senior civil design classes.

Dr. Bloetscher has also been the president of Public Utility Management and Planning Services Inc. (PUMPS) since he created it in July 2000. PUMPS is a consulting firm dedicated to evaluation of utility systems, needs assessments, condition assessments, strategic planning, capital improvement planning, grant and loan acquisition, inter-local agreement recommendations, bond document preparation, consultant coordination, permitting, and implementation of capital improvement construction. Prior to starting PUMPS, Dr. Bloetscher worked for 20 years in local government, holding positions including utilities engineer, deputy utility director, utility director, and city manager.

Dr. Bloetscher has also taught engineering at the University of Miami in Coral Gables, Fla., and participated in classes for local officials with the School of Government at the University of North Carolina at Chapel Hill. He is currently chair of the Water Resources Division Board of Trustees for the American Water Works Association (AWWA). He is the past chair of the AWWA Groundwater Committee and its Aquifer Storage and Recovery subcommittee. Dr. Bloetscher can be contacted at P.O. Box 221890, Hollywood, FL, 33022-1890 or h2o_man@bellsouth.net.

This page intentionally blank.